Indications and techniques of percutaneous procedures

Coronary, peripheral, and
structural heart disease

Indications and techniques of percutaneous procedures

Coronary, peripheral, and structural heart disease

Anthony A Bavry
University of Florida,
Gainesville, FL

Dharam J Kumbhani
Brigham and Women's Hospital,
Harvard Medical School,
Boston, MA

 Springer Healthcare

Published by Springer Healthcare Ltd, 236 Gray's Inn Road, London, WC1X 8HB, UK.

www.springerhealthcare.com

©2012 Springer Healthcare, a part of Springer Science+Business Media.

British Library Cataloguing-in-Publication Data.

A catalogue record for this book is available from the British Library.

ISBN 1 978-1-907673-18-4

Project editor: Anne Carty and Tamsin Curtis
Designer: Joe Harvey
Artworker: Sissan Mollerfors
Production: Marina Maher

Contents

Author biographies

Anthony A Bavry is Assistant Professor of Medicine at the University of Florida – Gainesville, FL and Director of the Veteran Affairs Medical Center Cardiac Catheterization Laboratory. Dr Bavry received his M.D. degree from the University of Florida, with honors in research, and also an M.P.H. degree focusing on Quantitative Methods from Harvard University. His internal medicine residency was at the University of Arizona where he also served as Chief Medicine Resident. He then completed a fellowship in cardiovascular medicine and interventional cardiology at the Cleveland Clinic with specialized training in peripheral vascular and structural heart disease and also was Chief Interventional Cardiology Fellow.

Dr Bavry's career interests are acute presentations and prevention of vascular disease. He has received several awards including the Florida Heart Research – Stop Heart Disease Researcher of the Year Award in 2009. He has published numerous original outcomes research manuscripts in peer-reviewed literature, edited several textbooks, and is a team leader for clinical trial summaries for the American College of Cardiology's CardioSource.

Dharam J Kumbhani is currently Instructor in Medicine, and Interventional Cardiology Fellow at Brigham and Women's Hospital and Harvard Medical School, Boston, MA. Dr Kumbhani received his M.D. degree with honors from Grant Medical College and Sir J. J. Group of Hospitals through the University of Mumbai. He received his M.S. degree in Epidemiology from Harvard University. He completed his internal medicine internship and residency at the University of Pennsylvania, and then a fellowship in cardiovascular medicine at the Cleveland Clinic.

Dr Kumbhani's research interests include outcomes research in acute coronary syndromes, and following structural heart disease and peripheral interventions. He has received numerous research accolades and was the runner up for the American College of Cardiology's Young Investigator Award in 2010. He has been nominated to the American College of

Epidemiology and the Sigma Xi Research Society. He has authored or co-authored several peer-reviewed articles, including in the *American Journal of Cardiology, American Heart Journal, American Journal of Medicine, European Heart Journal, Journal of the American College of Cardiology,* and the *Journal of the American Medical Association.* Dr Kumbhani is a peer reviewer for a number of prestigious journals including the *Annals of Internal Medicine.* He is a clinical trials team leader and member of the editorial board for the American College of Cardiology's Cardiosource. He is also an Associate Faculty Member of the Faculty of 1000 Medicine.

Disclaimer

This book is merely intended as a guide for performing percutaneous procedures. Proficiency in performing these procedures should be achieved through an accredited training program or by adequate proctoring. Complete details on the use of a device should always be obtained by referring to the product's indication for use.

Acknowledgments

We would like to thank Dr Samir R Kapadia, Director of the Sonnes Catherization Laboratories at Cleveland Clinic, for his review and much appreciated comments on this book. We would also like to thank Marion Tomasko and Joseph Pangrace from the medical illustration department at Cleveland Clinic for creating many of the superbly detailed illustrations included in this book.

Dedication

To my wonderful parents. Without your unending love and support, I could not have accomplished my goals.
Anthony A Bavry, MD, MPH

To my lovely wife, Meghna, and darling son, Rishi – for your love, support and sacrifice. To my wonderful parents – for everything that I am, and will ever be. To SB – for always inspiring me.
Dharam J Kumbhani, MD, SM

Abbreviations

ACA	anterior cerebral artery
ACT	activated clotting time
Ao	aortic
ASSD	atrial septal defect
AVR	aortic valve replacement
CAS	carotid artery stenting
CCA	common carotid artery
CEA	carotid endarterectomy
EPD	embolic protection device
FEP	fluorinated ethylene propylene
FFR	fractional flow reserve
Fr	French
HOCM	hypertrophic obstructive cardiomyopathy
IABP	intra-aortic balloon pump
ICA	internal carotid artery
IMA	inferior mesenteric artery
IVC	inferior vena cava
IVUS	intravascular ultrasound
LA	left arterial
LAO	left anterior oblique
LV	left ventricle
MCA	middle cerebral artery
MLA	minimal luminal area
N	nitinol
PFO	patent foramen ovale
PTFE	polytetrafluoroethylene
PU	polyurethane
RAO	right anterior oblique
RFA	right femoral artery
SMA	superior mesenteric artery
SVC	superior vena cava
TAVI	transcather aortic valve implantation
TEE	transesophageal echocardiography

Chapter 1

Vascular anatomy

Ascending aorta

The ascending aorta commences at the the base of the left ventricle. It is <3 cm in diameter and is usually situated about 6 cm posterior to the sternum. The right and left coronary arteries originate from the right and left sinus of Valsalva, respectively.

Aortic arch (transverse aorta)

The aortic arch begins at the level of the upper border of the right second sternocostal joint. It traverses in a superior–posterior direction initially, to the left and in front of the trachea, after which it is directed more posteriorly to the left of the trachea. It then passes inferiorly on the left side of the fourth thoracic vertebra (T4) to continue as the descending aorta. Its upper border is usually about 2.5 cm below the superior border of the manubrium. The three branches of the aortic arch are the innominate, left common carotid and left subclavian arteries. This pattern of branching is noted in approximately 70% of normal subjects. See Figure 1.1 for a diagram illustrating the ascending aorta, arch, their branches and relationship to surrounding structures.

The most common variant to this pattern is the so-called "bovine arch" (a misnomer), where there is a common origin of the innominate and left common carotid arteries. Infrequently, the left common carotid artery (CCA) may arise from the innominate artery. This is also sometimes referred to as a "bovine arch." This pattern of branching is noted in about 13% of normal subjects. (A true bovine arch, seen in cattle, has a single great vessel coming off the aortic arch, which gives rise to both

Ascending aorta, arch, their branches and relationship to surrounding structures

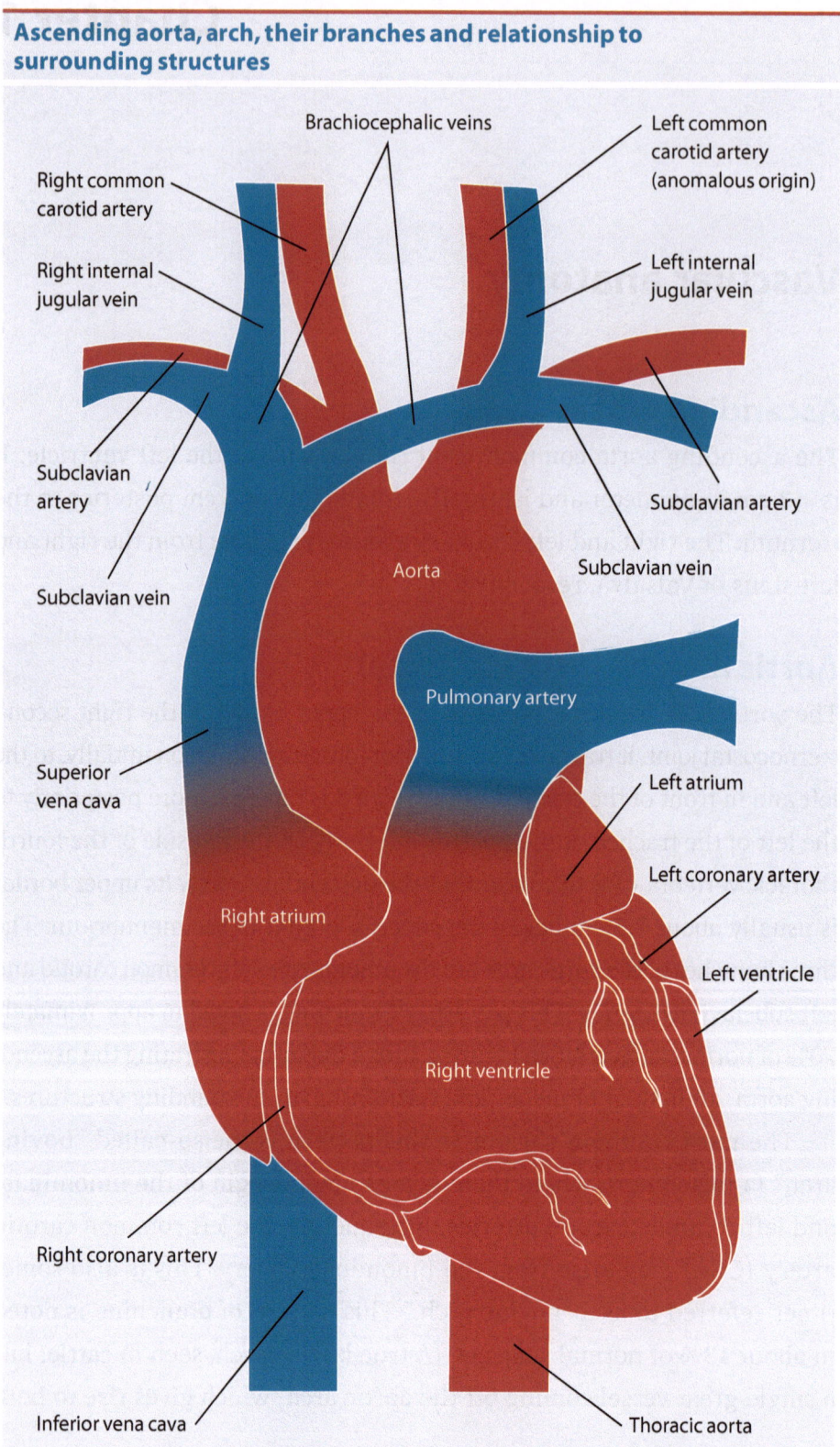

Figure 1.1 Ascending aorta, arch, their branches and relationship to surrounding structures.

subclavian arteries and a bicarotid trunk.) See Figure 1.2 for a diagram showing normal and "bovine" arch structures.

Sometimes there can be four branches arising from the aortic arch. Either the right subclavian and right carotid arteries have separate origins (with the innominate artery being absent), or the left vertebral artery comes off from the arch in between the left carotid and subclavian arteries.

From an interventional perspective, especially for carotid artery interventions, three different patterns of origin of the three neck vessels have been described, as demonstrated in Figures 1.2 and 1.3.

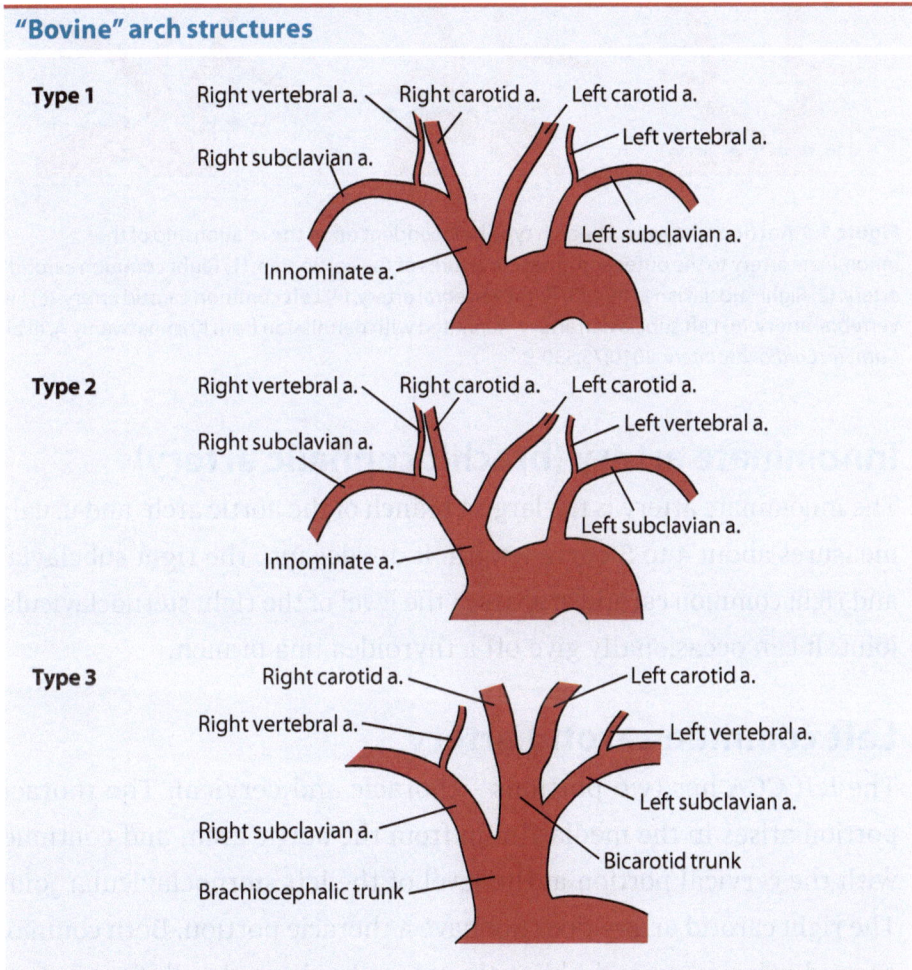

"Bovine" arch structures

Figure 1.2 "Bovine" arch structures. Type 1: The second most common pattern of human aortic arch branching has a common origin for the innominate and left common carotid arteries. This pattern has erroneously been referred to as a "bovine arch". Type 2: In this variant of aortic arch branching, the left common carotid artery originates from the innominate artery. This pattern has also been erroneously referred to as a "bovine" arch. Type 3: The aortic arch branching pattern found in cattle has a single brachiocephalic trunk originating from the aortic arch and eventually splits into the bilateral subclavian arteries and a bi-carotid trunk. Reprinted with permission from Layton KF, et al. *Am J Neuroradiol.* 2006;1541-2.

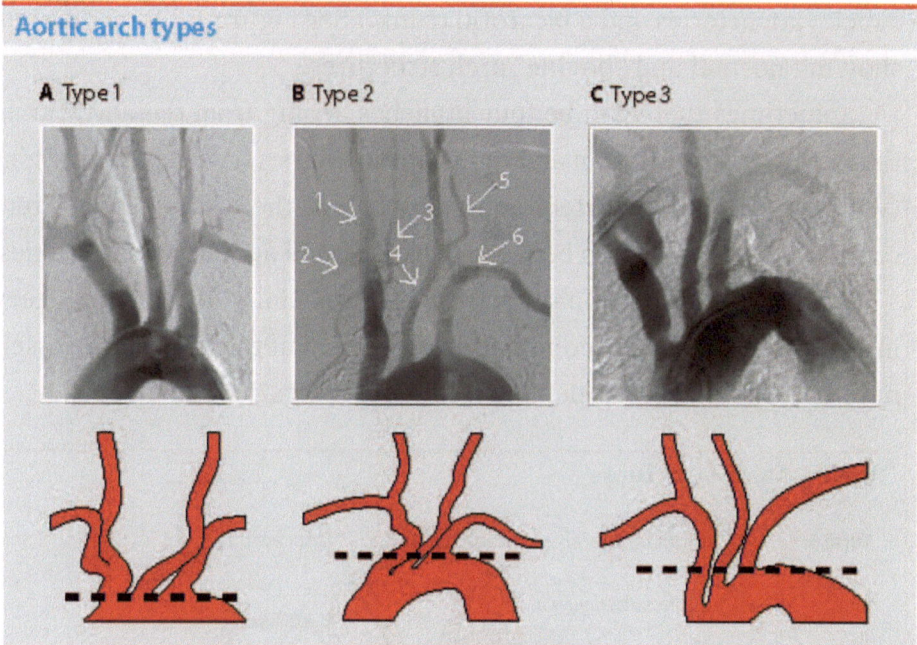

Aortic arch types

Figure 1.3 Aortic arch types. The arch type is dependent upon the relationship of the innominate artery to the outer and inner curvatures of the aortic arch. (1) Right common carotid artery. (2) Right subclavian artery. (3) Right vertebral artery. (4) Left common carotid artery. (5) Left vertebral artery. (6) Left subclavian artery. Reprinted with permission from Krishnaswamy A, et al. *Catheter Cardiovasc Interv.* 2010;75:530-9.

Innominate artery (brachiocephalic artery)

The innominate artery is the largest branch of the aortic arch, and usually measures about 4 to 5 cm in length. It divides into the right subclavian and right common carotid arteries at the level of the right sternoclavicular joint. It can occasionally give off a thyroidea ima branch.

Left common carotid artery

The left CCA has two portions – thoracic and cervical. The thoracic portion arises in the mediastinum from the aortic arch, and continues with the cervical portion at the level of the left sternoclavicular joint. The right carotid artery does not have a thoracic portion. Both common carotid arteries ascend obliquely upward, where they bifurcate into the external and internal carotid arteries at the level of the thyroid cartilage. See Tables 1.1, 1.2 and 1.3 for the branches and subdivisions of the internal and external carotid arteries. For a diagram showing the lateral projection of the left carotid artery see Figure 1.4.

Branches of the external carotid artery

Anterior	Superior thyroid
	Lingual
	External maxillary
Posterior	Occipital
	Posterior auricular
Ascending	Ascending
	Pharyngeal
Terminal	Superficial temporal
	Internal maxillary

Table 1.1 Branches of the external carotid artery.

Subdivisions and respective branches of the internal carotid artery

Cervical portion	None
Petrous portion	Caroticotympanic
	Artery to pterygoid canal
	Cavernous
	Hypophyseal
Cavernous portion	Semilunar
	Anterior meningeal
	Ophthalmic
Cerebral/supraclinoid portion	Anterior cerebral
	Middle cerebral
	Posterior communicating
	Choroidal

Table 1.2 Subdivisions and respective branches of the internal carotid artery.

Branches of the abdominal aorta

Visceral	Celiac
	Superior mesenteric artery (SMA)
	Inferior mesenteric artery (IMA)
	Middle suprarenals
	Renals
	Internal spermatic
	Ovarian (female)
	Testicular (male)
Parietal branches	Inferior phrenics
	Lumbars
	Middle sacral
Terminal branches	Common iliacs

Table 1.3 Branches of the abdominal aorta.

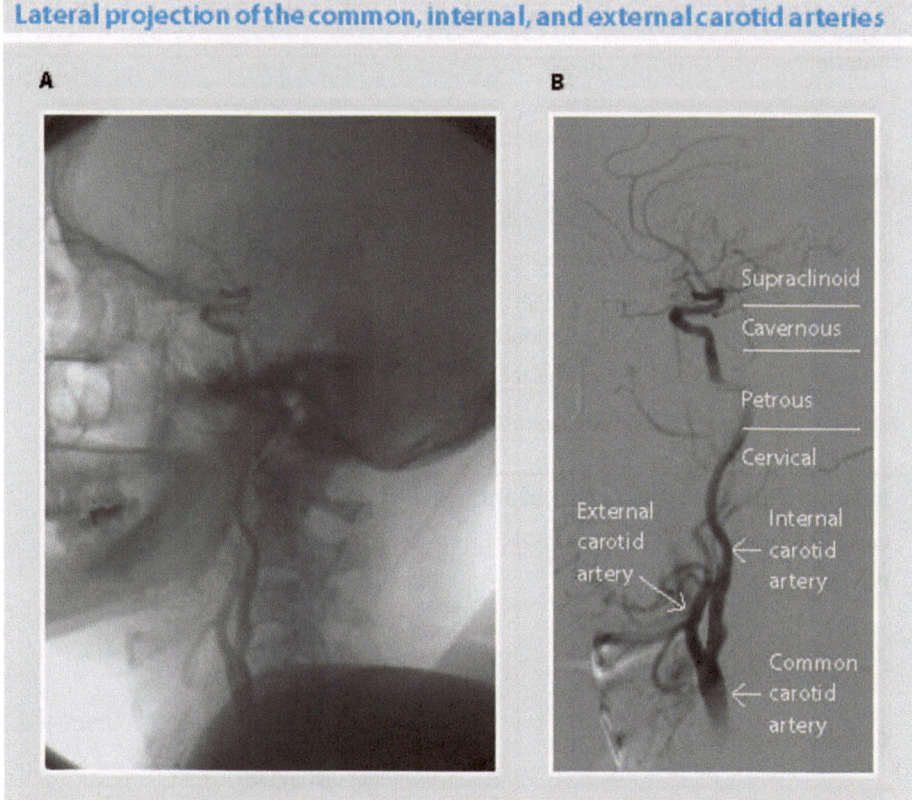

Figure 1.4 Lateral projection of the common, internal, and external carotid arteries. The individual segments of the internal carotid artery are labeled. The carotid bifurcation is typically at the angle of the jaw, but in this patient there is a low bifurcation. Reprinted with permission from Krishnaswamy A, et al. *Catheter Cardiovasc Interv.* 2010;75:530-9.

Left subclavian artery

The left subclavian artery is divided into three portions by the scaleneus anterior muscle. The first portion gives off all four branches of the left subclavian artery – vertebral, internal mammary, costocervical and thyrocervical arteries. Of note, each internal mammary artery arises about 2 cm above the respective clavicle's sternal end, opposite the root of the thyrocervical trunk.

Descending aorta

The descending aorta is divided into the thoracic and abdominal aorta. The thoracic aorta begins at the lower level of T4, to the left of the vertebral column, approaching the midline as it descends, and is anterior to it where it terminates into the abdominal aorta. It provides visceral branches to the pericardium, lungs, bronchi, and esophagus, and parietal branches to the thoracic wall. The abdominal aorta begins anterior to the inferior border of T12. It then descends down to end at the level of the fourth lumbar vertebra (L4) where it bifurcates into the left and right common iliac arteries, at a mean angle of approximately 37°. It follows the curvature of the lumbar vertebrae, being convex anteriorly. The peak of this convexity is at the level of L3. As it descends down, it narrows in diameter as many large branches come off it. The inferior vena cava (IVC) runs parallel to it on the right (Figure 1.5).

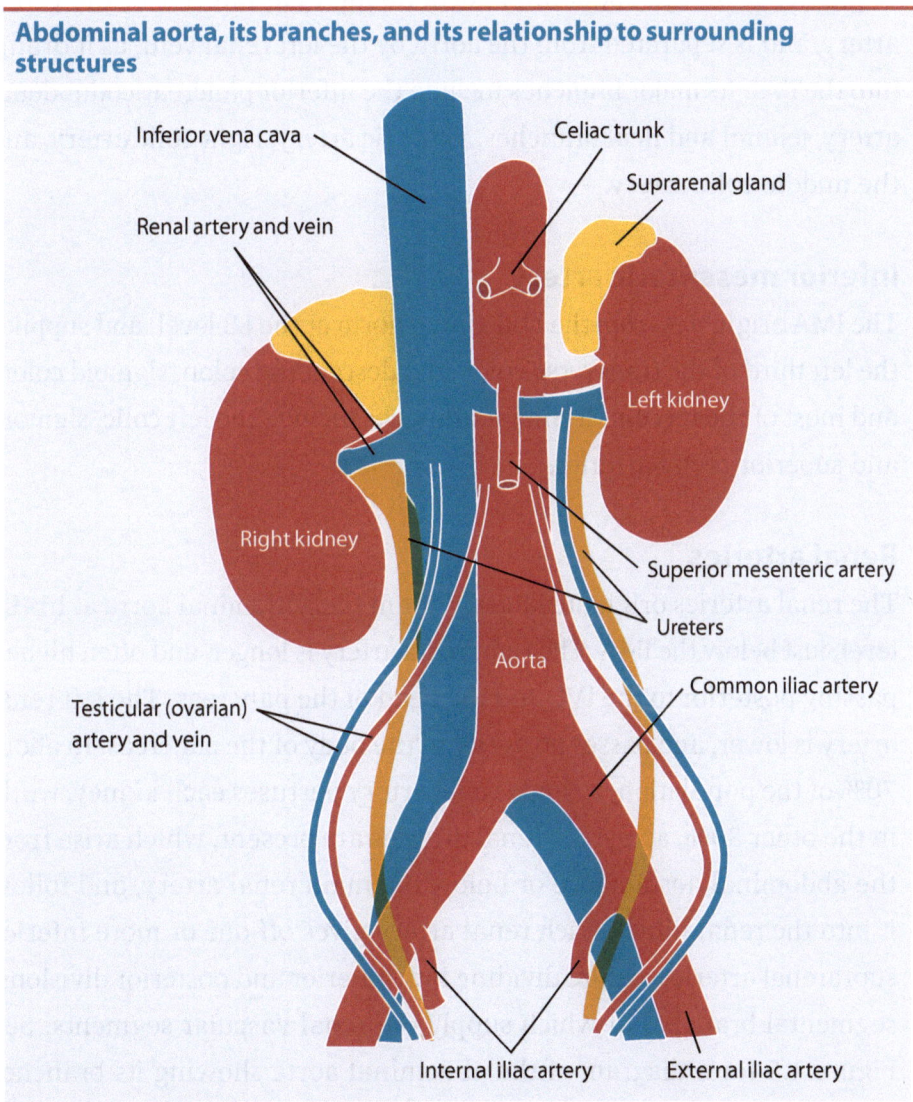

Abdominal aorta, its branches, and its relationship to surrounding structures

Inferior vena cava

Celiac trunk

Suprarenal gland

Renal artery and vein

Left kidney

Right kidney

Superior mesenteric artery

Ureters

Aorta

Common iliac artery

Testicular (ovarian) artery and vein

Internal iliac artery

External iliac artery

Figure 1.5 Abdominal aorta, its branches, and its relationship to surrounding structures.

Branches of the abdominal aorta

Celiac trunk

The celiac trunk comes off from the anterior aorta at the T12 level. It is about 1.25 cm long, and passes almost horizontally forward and slightly inferiorly before dividing into the splenic, common hepatic, and left grastric branches.

Superior mesenteric artery

The SMA arises at the L1 level, and supplies the majority of the small intestine, cecum, the ascending, and most of the transverse colon. It originates from the abdominal aorta about 1 cm inferior to the celiac artery, and is separated from the aorta by the left renal vein, as it drains into the IVC. Its major branches include the inferior pancreaticoduodenal artery, jejunal and ileal branches, ileocolic artery, right colic artery, and the middle colic artery.

Inferior mesenteric artery

The IMA originates from the abdominal aorta at the L3 level, and supplies the left third of the transverse colon, the descending colon, sigmoid colon, and most of the rectum. Its major branches include the left colic, sigmoid and superior rectal arteries.

Renal arteries

The renal arteries originate laterally from the abdominal aorta at L1–L2 level, just below the IMA. The right renal artery is longer, and often higher, passing posterior to the IVC and the head of the pancreas. The left renal artery is lower, and passes posterior to the body of the pancreas. In about 70% of the population, a single renal artery perfuses each kidney, while in the other 30%, accessory renal arteries are present, which arise from the abdominal aorta above or below the main renal artery, and follow it into the renal hilum. Each renal artery gives off one or more inferior suprarenal arteries, before dividing into anterior and posterior divisions, segmental branches of which supply the renal vascular segments. See Figure 1.5 for a diagram of the abdominal aorta showing its branches and its relationship to surrounding structures.

Chapter 2

Catheters, guidewires, sheaths, and balloons

Catheters are workhorse devices that are used every day in interventional cardiology. They are broadly grouped into three categories: diagnostic catheters, guide catheters, and micro or support catheters. In addition to the role of each group and type of catheter, it is also important to have a good working knowledge of the length of the various catheters, especially with chronic total occlusions and peripheral interventions. Commonly used catheters are shown in Figure 2.1.

Diagnostic catheters

Diagnostic catheters are flexible devices that are relatively resistant to kinking. They are sized by their external diameter using the French (Fr) system. This size also provides a crude reference diameter of the vessel to be stented. The outer diameter, in mm, of the catheter is determined by dividing the Fr size by 3. For example, a 5 Fr catheter has a diameter of 1.67 mm, a 6 Fr 2.0 mm, a 7 Fr 2.3 mm, 8 Fr 2.7 mm, and so on. The standard length of a diagnostic catheter is 100 cm. This provides approximately 15 cm of working catheter outside the body. Long diagnostic catheters are 125 cm and are frequently used for peripheral angiography and interventions. Examples include the Judkins right (JR), multipurpose (MP), Headhunter (H) slip, JB1 slip, and glide catheters. Some of these catheters have a hydrophilic coating on their surface that facilitates advancement. Specialized diagnostic catheters for carotid angiography are the Vitek

Commonly used diagnostic catheters

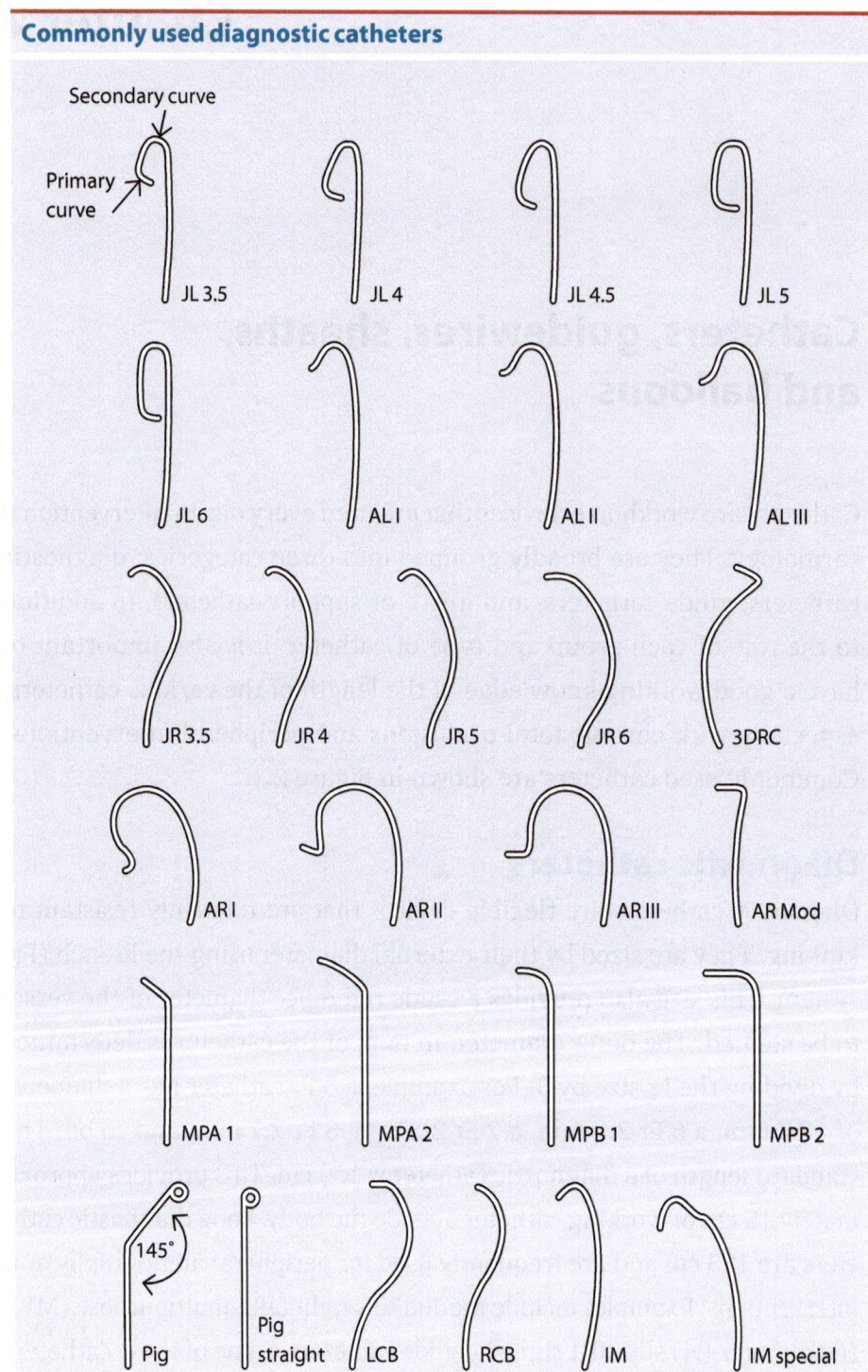

Figure 2.1 Commonly used diagnostic catheters. AR, Amplatz right; AR Mod, Amplatz right modified; AL, Amplatz left; IM, internal mammary; JL, Judkins left; JR, Judkins right; LCB, left coronary bypass; MPA, multipurpose A; MPB, multipurpose B; Pig, pigtail; RCB, right coronary bypass. Reprinted with permission from Aviles RJ, et al. *Introductory Guide to Cardiac Catheterization*. Philadelphia, PA: Lippincott Williams & Wilkins, 2004:31.

Settings for large volume left heart angiography

	Ascending aorta	Aortic arch	Distal aorta	Abdominal aortogram with lower extremity run-off	Iliac
Rate	15–20 cc	15–20 cc	15–20 cc	15 cc	(7) 10 cc
Volume	30–40 cc (60 cc to assess aortic root or unknown saphenous vein graft anatomy)	30–40 cc	30–40 cc	90 cc	(14) 20 cc
Catheter	Angled Pig	Angled Pig	Straight Pig, angled Pig, or straight flush	Straight Pig, angled Pig, or straight flush	Straight Pig, angled Pig, or straight flush

Table 2.1 Settings for large volume left heart angiography. Pig, pigtail catheter.

Settings for right heart angiography

	Pulmonary artery	RA/IVC
Rate	20–40 cc	20 cc
Volume	40 cc	30–40 cc
Catheter	Berman, NIH, Pig	NIH

Table 2.2 Settings for right heart angiography. IVC, inferior vena cava; RA, right atrium.

and Simmons catheters. Commonly used settings for routine right and left heart catheterization are included in Tables 2.1 and 2.2.

Guiding catheters

Guide catheters are corrugated with metal fibers throughout their length, which allows for a reduction in the catheter material and an increase in the internal diameter. Therefore, the internal diameter of a guide catheter is greater than a diagnostic catheter for the same Fr size. The design of the guide catheter provides more support than diagnostic catheters. Similar to diagnostic catheters, the standard length is 100 cm. Some guide catheters are available in 90-cm length. These are used in special circumstances; for example, for a retrograde chronic total occlusion intervention, such as a chronic total occlusion of the right coronary artery that is attempted to be opened by a retrograde approach from the left anterior descending through a septal perforator. If the lesion is in the proximal right coronary artery, a standard balloon with a 120-cm shaft inserted through a 100-cm guide would not reach the lesion, whereas a shorter guide catheter (i.e., 90 cm) and a 120-cm balloon shaft would allow the balloon to reach the proximal right coronary artery lesion.

Guiding sheaths

Guiding sheaths take the place of a guiding catheter and access sheath system. It is a combined device that obviates the need for a separate access sheath. The guiding sheath is straight or close to straight; therefore, it has little direction, unlike the guiding catheters, which have directionality (i.e., JR4 guide catheter). Diagnostic catheters can be telescoped out of the end of the guiding sheath. For example, a diagnostic catheter (i.e., 5 Fr JR4) is used to engage the artery of interest, and the guiding sheath is railed over it into position. Guiding sheaths are not as stiff as guiding catheters, so they are malleable and can track distal into an artery. For carotid intervention, a guiding sheath is advanced all the way to the distal (CCA). Examples of guiding sheaths are the Ansel, Raabe, and Shuttle™ sheaths. Figure 2.2 shows a Shuttle sheath in a subclavian artery intervention.

Microcatheters or support catheters

Micro or support catheters serve various purposes. An over-the-wire balloon is actually a support catheter in that it provides an additional degree of support throughout the system to enable a wire to cross a difficult lesion. Progreat™, Transit, and the Excelsior® are examples of dedicated support catheters. See Table 2.3 for a summary of guiding catheters and sheaths for peripheral interventions.

Special circumstances

Occasionally, patients with coronary artery bypass grafting will have had a gastroepiploic artery used as a conduit to the distal right coronary artery. A JR4 catheter can initially be used to engage the celiac trunk; however, this catheter may not have sufficient reach. Alternative catheters include the Amplatz right 1 or 2, Cobra, SOS, Vitek, and the Simmons 1, 2, or 3.

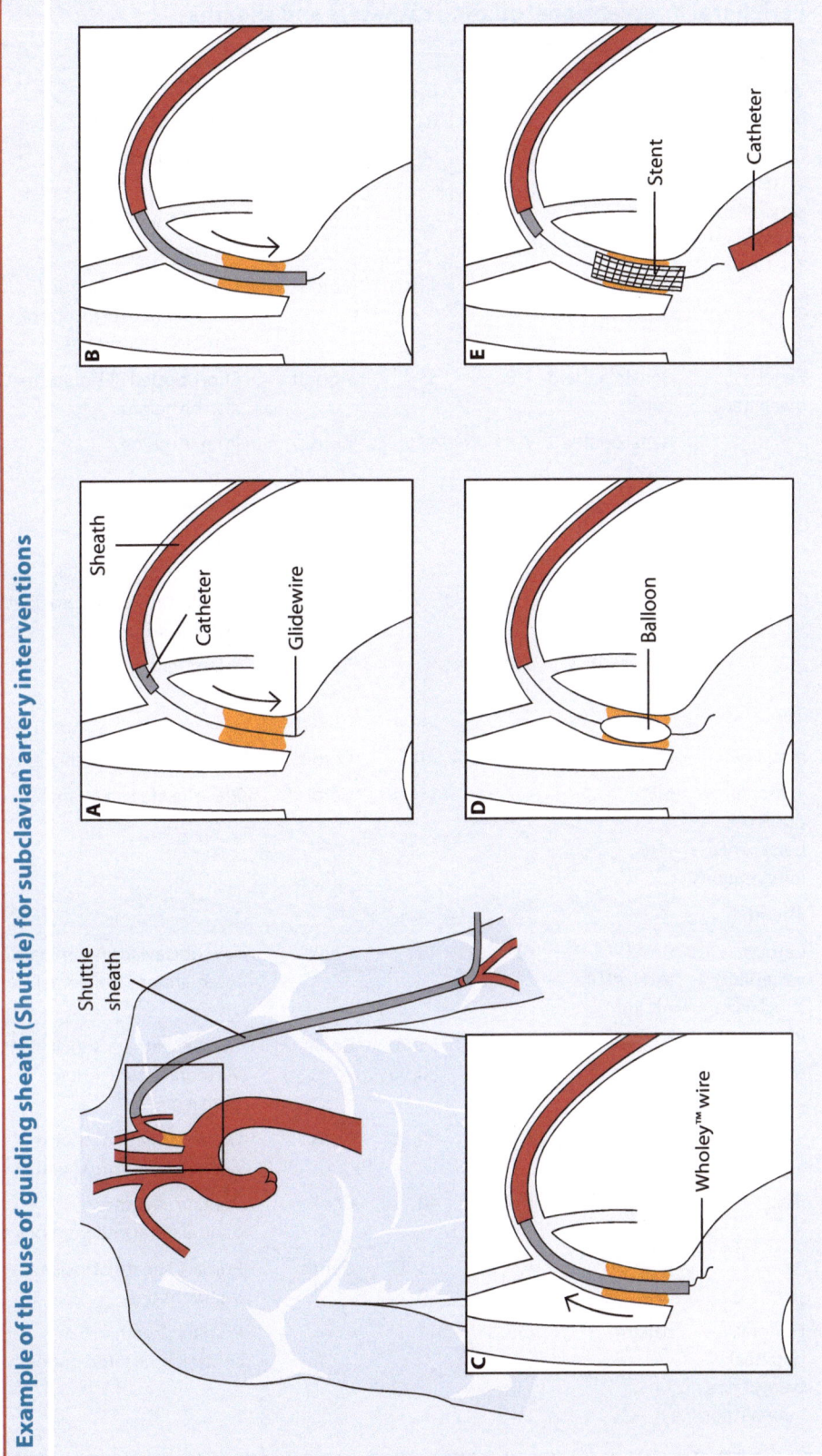

Figure 2.2 Example of the use of guiding sheath (Shuttle) for subclavian artery interventions. Reprinted with permission from Bhatt DL. *Guide to Peripheral and Cerebrovascular Intervention*. London, UK: Remedica Publishing, 2004:123-49.

Peripheral interventional guiding catheters and sheaths

Catheters

Vascular bed	Name	Size (French)	Length (cm)	Company	Comments
Carotid, vertebral, subclavian, innominate	Headhunter (H1)	8, 9	90	Mayo Healthcare	Primary guide
	Judkins right (JR 4)	8, 9	110	Cordis	Good for vertebral and subclavian arteries
	Amplatz left (AL I)	8, 9	110	Cordis	Useful for type 3 aortic arch, marked tortuosity of common carotid artery
Renal/ mesenteric	Renal standard curve	7, 8	55	Abbott	Short-tipped RES guide may also be helpful
	Renal double curve	7, 8	55	Cordis/ Abbott	Primary guide
	Renal multi-purpose (MP)	7, 8	55	Cordis	For arm cases
	JR 4	7, 8	110	Cordis	Good for small, heavily calcified aorta, but provides little back-up
	Hockey Stick	7, 8	55	Boston Scientific/ Guidant	Infrequent use
Iliac	MP	7, 8	110	Cordis	If access is from brachial artery
Femoral, popliteal, below knee interventions	MP	6, 7, 8	110	Cordis	Place inside Balkan sheath

Sheaths

Vascular bed	Name	Size (French)	Length (cm)	Company	Comments
Carotid, vertebral, subclavian, innominate	Ansel (AN I, AN II, AN III)	6, 7	45	Cook	For subclavian/innominate interventions from brachial artery
	Raabe	6, 7, 8, 9	55, 70, 80, 90	Cook	Good sheath for subclavian/ vertebral interventions from leg
	Shuttle sheath	6, 7, 8, 9	80, 90	Cook	Carotid sheath of choice Excellent tracking sheath
Iliac	Balkin®	6, 7, 8	40	Cook	Primary sheath for contralateral interventions
	Brite Tip®	6, 7, 8	35, 55, 90	Cordis	Primary sheath of ipsilateral interventions
Femoral, popliteal, below knee interventions	Balkin	6, 7, 8	40	Cook	Primary sheath for contralateral interventions

Table 2.3 Peripheral interventional guiding catheters and sheaths. Adapted from Casserly IP, et al. *Manual of Peripheral Intervention*. Philadelphia, PA: Lippincott Williams & Wilkins, 2005.

Multipurpose catheters come in two shapes referred to as A and B with a choice of side-holes (i.e., '1' for one end-hole and '2' for two side-holes). The A curve is approximately 45°, while the B curve is approximately 90°.

Profiles of commonly used wires

Size	Indication		
0.009"	Rotational atherectomy wire		
0.014"	Coronary wire		
0.018"	Micropuncture wire		
	Precise® self-expandable stent		
	Savvy® balloon		
	Coil deployment		
0.025"	Swan-Ganz catheter		
	Intra-aortic balloon pump (7.5 and 8 Fr)		
0.030	Intra-aortic balloon pump (9.5 and 10.5 Fr)		
0.032"	Mullins sheath for transseptal puncture		
0.035"	Short wire in kit for access sheath (≥4 Fr)		
	Advancing diagnostic and interventional coronary catheters		
0.038"	Short wire in kit for access sheath insertion (≥5 Fr)		
Situation	**Ideal property of wire**	**Commonly used wires**	**Company**
Regular intervention	Flexible atraumatic tip, with low to moderate support	Balance Trek™	Guidant
		Balance Middleweight™	Guidant
		Asahi Light	Abbott Vascular
		Asahi Soft	Abbott Vascular
		CholCE® Floppy	Boston Scientific
		Hi-Torque Floppy	Boston Scientific
		Wizdom Floppy	Cordis
		Asahi Prowater	Abbott Vascular
Tortuosity	Steering and tracking more important than support. Tapered core-to-tip design, with hydrophilic coating or hydrophilic plastic polymer	Whisper	Guidant
		PT Graphix™	Boston Scientific
		CholCE PT	Boston Scientific
		PT2®	Boston Scientific
Significant tortuosity, severe proximal angulations	Extra support needed. Atraumatic flexible tip with a large stainless-steel core	Stabilizer™	Cordis
		Ironman™	Guidant
		Balance Heavyweight™	Guidant
		Mailman™	Boston Scientific
Chronic total occlusions	Wire tip core should have some body, steerability important. Tip-to-core construction, shorter steep distal core taper, tapered tip	Cross-IT	Guidant
		Asahi Miracle Bros	Abbott Vascular
		Confianza	Abbott Vascular
		Hydrophilic PT Graphix	Boston Scientific
		Shinobi™	Cordis

Table 2.4 Profiles of commonly used wires.

Anterior take-off of the right coronary artery can be engaged with either a Williams Right or a 3DRC (No-Torque Right) catheter, or alternatively, with one of the Amplatz Left (AL) catheters such as AL 0.75 or AL I.

Profiles of commonly used wires can be found in Table 2.4.

Balloons

Aviator™ and Viatrac™ are examples of balloons that track rapid exchange over a 0.014" wire. The largest diameter of these balloons is 7 mm. The need for a larger balloon will require a 0.035" over-the-wire system (e.g., OPTA® Pro or Agiltrac® balloons), with maximal diameter of 12 mm. Peripheral stents come on one of the above mentioned balloons; for example, a Genesis™ on an Aviator with a maximal diameter of 7 mm, or a Genesis on an OPTA Pro with a maximal diameter of 12 mm (Table 2.5).

Peripheral balloon and stents			
Maximal diameter wire	**0.0014"**	**0.018"**	**0.035"**
Balloons	Aviator	Savvy	OPTA Pro
	Viatrac		
	Savvy		
Stents	Acculink™	Precise	Genesis
	Xact®		Palmaz® Schatz/Blue
	Protege®		

Table 2.5 Peripheral balloon and stents.

Chapter 3

Complex coronary interventions

Rotational atherectomy

> **Definition**
> Removal or controlled scoring of the obstructing calcified athero-sclerotic plaque (rather than mere displacement).

Indications

1. Severely calcified lesion that prevents complete balloon expansion during angioplasty.
2. Severely calcified lesion that is able to be crossed with a coronary wire, although unable to cross with a balloon or stent.
3. Severe ostial lesions.

Access

Right or left femoral artery.

Anticoagulation

Aspirin, clopidogrel, and heparin. Heparin is preferred over bivalirudin in case rapid reversal of anticoagulation is needed in the event of a perforation.

Equipment

1. Short 6–8 Fr sheath.
2. 6–8 Fr coronary guide catheter.

3. RotaWire® floppy wire (0.009"), 300 cm; infrequently RotaWire extra support wire (0.009").
4. Rotalink® burr (smallest size is 1.25 mm).
5. Rotalink advancer system (advancer, flush, pedal).
6. Aminophylline.
7. Possibly temporary pacer for right coronary artery (RCA) lesions.

Procedure

1. Place a temporary pacemaker in the right ventricular apex if indicated.
2. Cross the lesion with the wire. This is usually done with the RotaWire floppy wire, although this wire is somewhat difficult to manipulate as it is thin with poor support. (See pointers box at the end of this section for additional advice.) The standard 0.014" torquer (i.e., steering device) can be used on this wire or the WireClip® torquer that comes with the system.
3. Set up the Rotalink advancer system at the patient's feet. This may require an extra table beyond the catheterization table. Use a clip to fasten the black cable and pressure tubing from the advancer together.
4. Before the burr is attached to the rotational shaft, make sure the advancer knob is loosened. The microsheath on the burr shaft end needs to be pulled toward the burr end to expose the pin, which will attach to the rotational shaft. Attach the selected burr to the rotational shaft. Start with a small burr (typically 1.25 mm or 1.5 mm) and increase the size if necessary (in 0.5-mm increments). Note that the final burr-to-artery ratio should be 0.70–0.85. Advance the burr over the wire until the burr is close to the guide. Place the torquer on the wire at the back of the advancer unit. This serves as a secondary brake (the primary brake is in the advancer).
5. Make sure the Rotaglide® is running before advancing the burr into the guide. Set the rotational speed between 160,000 and 180,000 rpm and test outside the body.
6. This step requires two physicians. The first operator should be positioned at the back of the catheterization table and pin the wire and advance the Rota unit, while the other physician advances the

burr into the guide catheter. Place the burr in the coronary artery about 1 to 2 cm proximal to the intended lesion. Make sure the advancer knob is loose in the central position, so that the burr can be advanced and withdrawn by moving the knob. Aminophylline can be administered, especially for RCA lesions. When rotational atherectomy is ready to be performed, step on the pedal. Limit runs to 20–30 seconds with a 30-second or longer recovery period. The recovery period is important in case there are bradyarrhythmias, hypotension, or slow flow. Use either a slow forward pressure or a gentle pecking motion. The Rotablator® shaft tip can potentially break off or get stuck distally if it is forced through a lesion. Continue rotational atherectomy until there are no longer decelerations. This is usually about 2 to 3 passes per burr size.

7. Two people are needed for the removal of the burr, which is done on the DynaGlide™ mode (the button to the right of the foot pedal that controls the Rotablator System turbine and burr), which spins the burr at 60,000 rpm. The brake must be released throughout removal (black button on the back of the Rotalink advancer system). The first operator opens the hemostatic valve and the second operator walks the assembly back over the wire into the guide catheter to the level of the ascending aorta while keeping the wire fixed in place. Once the burr is in the guide catheter at the level of the ascending aorta or aortic arch, the second operator can let go of the wire and smoothly and steadily pull the Rotalink advancer system back until it reaches the torquer that was placed on the wire. Figure 3.1 (a–d) shows a Rotablator.

Profile of Rotalink® burr catheter

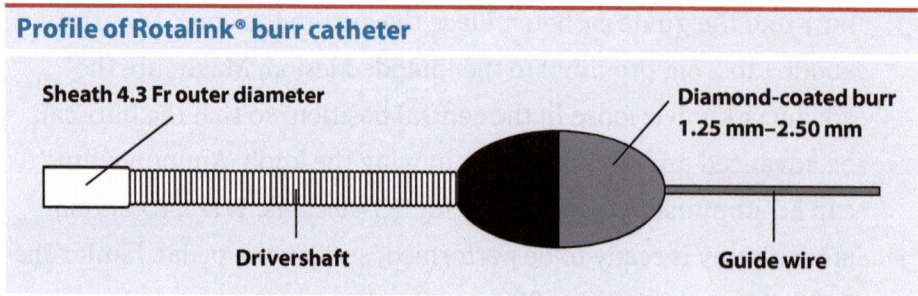

Figure 3.1a Profile of Rotalink® Burr Catheter. Proximal surface of the burr is diamond-tipped.

Rotalink advancer system

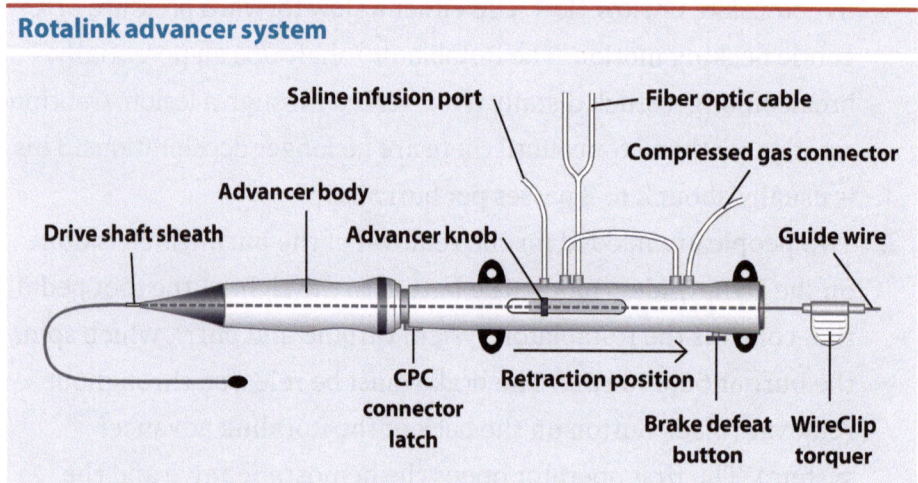

Figure 3.1b Rotalink advancer system. This device acts as a support for the air turbine and as a guide for the sliding elements, which control burr extension. Manipulation of the advancer knob allows independent extension of the burr.

RotaWires used for rotablation

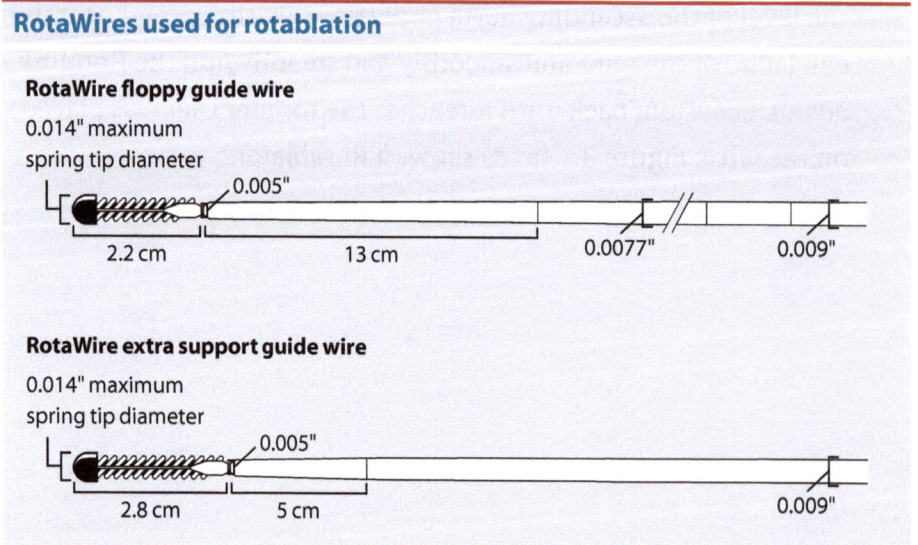

Figure 3.1c RotaWires used for rotablation.

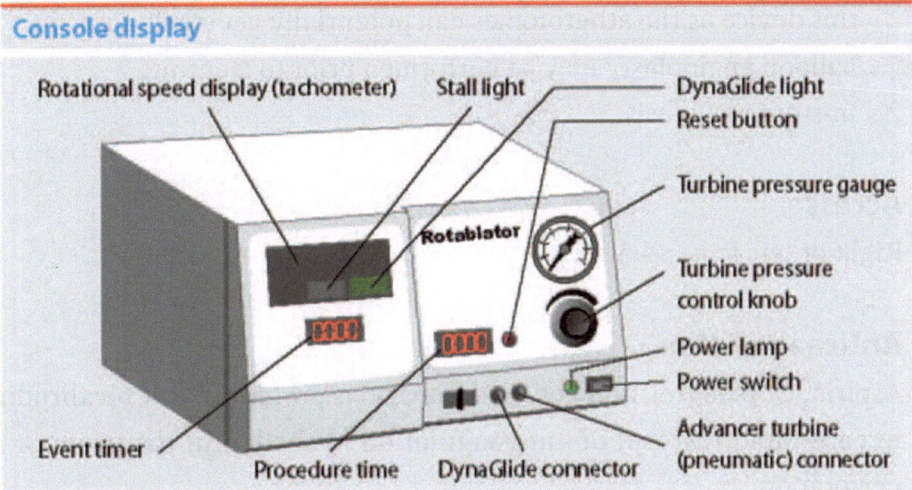

Figure 3.1d Console display.

Pointers

The RotaWire floppy wire is difficult to navigate since it has minimal support (0.009"). An easier approach is to cross the vessel with a standard coronary wire (0.014") and exchange it for a RotaWire floppy wire with a 1.5 mm over-the-wire balloon or a microcatheter (i.e., Transit catheter, or Progreat catheter). Unfortunately, the coronary lesion may be too severe to allow passage of the over-the-wire balloon or microcatheter. Alternatively, a RotaWire extra support wire may be used when more support is needed, such as distal lesions or heavily calcified proximal lesions.

Cutting balloon angioplasty

Definition

Use of a noncompliant balloon with longitudinally mounted microsurgical atherotomes to cause scoring of atheromatous plaque, and severing of elastic and fibrotic continuity of the vessel wall, resulting in plaque depression.

Indications

1. Ostial, bifurcation, or nonpassable lesions, which are only mildly calcified. Severe calcification is a contraindication to the use of

this device as the atherotomes can potentially get stuck. Cutting balloon angioplasty may be performed prior to stenting.

2. In-stent restenosis.

Access

Right or left femoral artery.

Anticoagulation

Aspirin, clopidogrel, and heparin. Heparin is preferred over bivalirudin in case rapid reversal of anticoagulation is needed in the event of a perforation.

Equipment

1. Short 6–8 Fr sheath.
2. 6–8 Fr coronary guide catheter, 100 cm.
3. Standard coronary 0.014" wire.

Procedure

1. Currently available cutting balloons include the AngioSculpt® and the Flextome®. Before the procedure, determine which one will be used and whether a 300-cm wire is required.
2. Cross the lesion with the wire.
3. Dilate the lesion with the cutting balloon.
4. Repeat as necessary.
5. Stent if indicated.

Pointers

Cutting balloon angioplasty may be intended as a stand alone procedure (i.e., without stenting); however, it is important to remember that iatrogenic dissections may occur, especially in native coronary arteres. Therefore, the operator should be prepared to perform stenting if necessary.

Complications

Cutting balloon angioplasty carries the same complications as with standard angioplasty; however, as described above, dissections may occur more frequently.

Figure 3.2 shows the profile of a cutting balloon.

Profile of cutting balloon

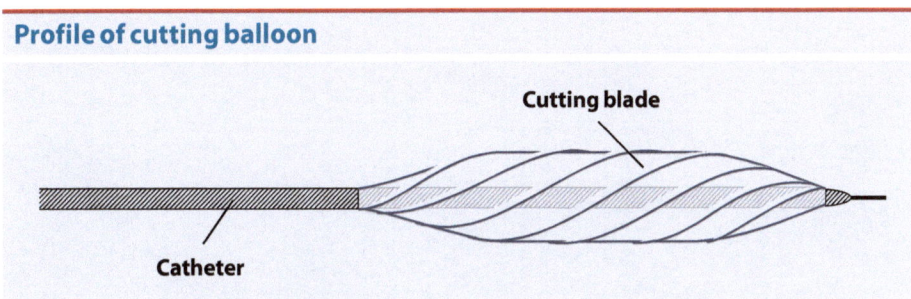

Cutting blade

Catheter

Figure 3.2 Profile of cutting balloon.

Chapter 4

Fractional flow reserve and intravascular ultrasound

Coronary angiography has inherent limitations. Fractional flow reserve (FFR) is a useful technique for determining the presence of ischemia with indeterminate lesions. Intravascular ultrasound (IVUS) is useful for optimizing stent implantation.

Fractional flow reserve

FFR is performed under maximal hyperemia (adenosine is given intravenously at 140 µg/kg/min). FFR <0.75 to 0.80 is considered flow-limiting and may therefore be appropriate for revascularization. In contrast, patients with FFR >0.80 do not appear to benefit from revascularization. Some operators consider a value between 0.75 and 0.80 to represent a "gray zone," where the decision to revascularize is often based on other supporting evidence, such as clinical presentation. Patients with non-flow limiting lesions who receive a stent do worse than medically treated patients (Table 4.1).

Once the decision is made to perform stenting, IVUS is an excellent tool to optimize stent implantation. IVUS should be used to determine the presence of calcium (i.e., need for rotational atherectomy), reference lumen diameter (i.e., for stent sizing), and stent expansion and apposition to the vessel wall after deployment. See Figure 4.1 for an example of IVUS imaging.

Considerations for fractional flow reserve (FFR)

FFR: Indications

- Angiographically borderline stenosis (i.e., 40–60%), especially if long/diffuse narrowing
- Multivessel PCI
- Bifurcation lesions when ostium of side branch looks compromised following stenting of main vessel

FFR: False negatives

- Failure to achieve maximal hyperemia (caffeine intake, compromised vasculature in patients with endothelial dysfunction, syndrome X, diabetes, severe hypertension, hypertrophic obstructive cardiomyopathy)
- Acute myocardial infarction
- Technical issues (catheter damping pressure resulting in false zero of pressure wire)
- Electronic/mechanical issues (failure to equalize properly, electronic drift)

Table 4.1 Considerations for fractional flow reserve (FFR).

Intracoronary IVUS images

Figure 4.1 Intracoronary IVUS images: A) Intravascular ultrasound (IVUS) catheter. B) Arterial lumen. C) Interface between intima and media. Note, the intima is only one cell layer thick (i.e., the endothelium). D) Atheromatous plaque within the media. E) External elastic membrane, which is the interface between the media and adventitia. F) Adventitia. The right panel highlights the minimal luminal area (MLA).

Intravascular ultrasound

Reference lumen diameter

IVUS is performed proximal and distal to the lesion to determine the reference lumen diameter that will be used for selecting the stent size. Note, that the size is determined from the reference lumen and not the lesion since the latter tends to over-size the stent from positive remodeling.

Glagov's phenomenon

The vessel lumen size is retained as the plaque volume increases. This is positive remodeling and it is accomplished by expansion of the external elastic media. Measurement of the external elastic media is used to determine the percentage of plaque volume and should not be used for sizing. This measurement should only be used in core labs for research purposes.

Stent cross-sectional area

After stent deployment, IVUS can be used to make sure the entire length of the stent is well expanded. A representative section is used to determine the stent cross-sectional area. Under-expanded stents should be further expanded with a noncompliant balloon.

Lumen cross-sectional area

The lumen cross-sectional area can be used as a surrogate measurement to assist the decision as to whether or not to stent a lesion. There are however difficulties with this measurement. For a non-left main trunk lesion, a lumen cross-sectional area of >4 mm² portends a good prognosis. Lesions with a cross-sectional area <4 mm² are generally considered flow-limiting. For left main trunk lesions, the cut-off is 6–7.5 mm². Since a small luminal cross-sectional area is nonspecific, we believe that the decision to stent should be based more on a significant FFR and less on the lumen cross-sectional area.

Chapter 5

Peripheral interventions

There are several different types of balloon- and self-extendable stents. Typical diameter sizes are summarized in Table 5.1. Guides are typically 100 cm long; short guides are 90 cm long. The standard length of coronary balloons is 110 cm. For peripheral interventions, the Savvy balloons come in two sizes: 135 and 150 cm; while the OPTA Pro balloons are either 110 cm or 135 cm long.

Renal artery interventions

Indication

Revascularization of severe unilateral renal artery disease has been studied in several clinical trials. Most have failed to document a benefit in

Typical sizes of stents for peripheral interventions	
Anatomic site	**Size (mm)**
Balloon-expandable stents	
Common iliac	7–10
External iliac	5–7
Renal artery	4–7
Subclavian	7–8
Self-expandable stents	
Internal carotid	6–8
Common carotid	8–10
SFA	7–8

Table 5.1 Typical sizes of stents for peripheral interventions. Note that self-expandable stents are usually oversized 1 to 2 mm relative to the reference vessel diameter. SFA, superficial femoral artery.

Current American College of Cardiology/American Heart Association indications for renal artery revascularization

1. Recurrent flash pulmonary edema in the setting of severe hypertension or unexplained recurrent congestive heart failure (Class I)

2. Severe renal artery stenosis in patients with accelerated, resistant or malignant hypertension, hypertension with intolerance to medications, and hypertension with an unexplained solitary kidney (Class IIa)

3. Progressive renal failure in patients with bilateral stenosis or severe stenosis in a solitary kidney (Class IIa)

4. Severe renal artery stenosis in patients with unstable angina (Class IIa)

5. Severe asymptomatic bilateral renal artery stenosis or in a solitary viable kidney (Class IIb)

Table 5.2 Current American College of Cardiology/American Heart Association indications for renal artery revascularization. Adapted from Hirsch AT, et al. *J Am Coll Cardiol.* 2006;1239–312.

this patient population. Current indications for renal artery vascularization are summarized in Table 5.2.

Angiogram

The renal arteries are often initially visualized nonselectively with an abdominal aortic angiogram. This can be performed through a variety of sheath sizes; however, the 5 Fr sheath is most typical. A straight pigtail catheter or straight flush catheter is advanced to the T12 level and power injection performed with 15–20 cc/sec of contrast for 30–40 cc.

Selective angiogram is performed with a 5 Fr JR 4 catheter, gently interrogating at the L1 to L2 level. The renal arteries are best visualized with 10° of ipsilateral angulation and 9" or lower magnification to get the entire kidney in the field of view.

Access

Right or left common femoral artery.

Anticoagulation

Aspirin, clopidogrel, and heparin.

Equipment

1. 7 Fr short sheath.
2. 55-cm renal guide (e.g., the Renal Short Standard [RESS] guide).

3. 150-cm soft-tipped 0.035" wire for advancing the catheter to the level of the renals.
4. Stiff 0.014" wire (Balance Heavyweight, Ironman, Grand Slam).

Intervention

There are three main techniques for performing renal artery revascularization: direct engagement, telescope technique, and no-touch technique. Intervention requires upsizing the access sheath to 6 or 7 Fr. Stenting is indicated in all patients who meet criteria for revascularization, except in patients with fibromuscular dysplasia where angioplasty alone is equally efficacious. Stenting should be used as an emergency procedure (e.g., if dissection occurs) in these patients.

Direct engagement

This technique directly engages the renal artery with the 6 or 7 Fr RESS or RES guide catheter. While this is the easiest technique, it has the most potential to damage the renal artery.

Telescope technique

A 5 Fr JR 4 catheter is telescoped out of a 7 Fr RESS or RES guide and advanced to the level of the renal arteries with a soft-tipped 0.035" wire. The wire is withdrawn and the JR 4 catheter is used to engage the renal artery. The diagnostic catheter is fixed in place and the guiding catheter is gently railed into the ostium of the renal artery.

No-touch technique

A 7 Fr RESS or RES guide is advanced over a soft-tipped 0.035" wire just under the level of the renal artery (L1 to L2). The 0.035" wire is left in place, and the renal artery is crossed with a 0.014" wire (typically a Balance Heavyweight or Ironman). At this point the 0.035" wire is withdrawn, keeping the guide catheter in the aorta, non-engaged in the renal artery.

Carotid artery interventions

Indication

Carotid artery stenting (CAS) is indicated for both symptomatic and asymptomatic patients who are at high surgical risk. In the future, carotid stenting may be used in lower surgical risk patients as well, although this still needs to be investigated in clinical studies.

For symptomatic patients, CAS is indicated with a 70% or greater stenosis, or a 50–69% stenosis and enrollment in a clinical trial or registry. CAS revascularization is also indicated in high-risk asymptomatic surgical patients enrolled in a clinical trial or registry with at least an 80% stenosis of the internal carotid artery (ICA). Current data do not support carotid stenting in asymptomatic patients with less than 80% stenosis or patients without high-risk features for surgical revascularization (see Tables 5.2, 5.3, and 5.4).

High-risk features for surgical carotid revascularization

Clinical

- Age ≥80 years
- Congestive heart failure with New York Heart Association Class III/IV symptoms
- Known severe left ventricular dysfunction, with ejection fraction ≤30%
- Open heart surgery necessary within 4–6 weeks
- Recent myocardial infarction (<4 weeks)
- Unstable angina (Canadian Cardiovascular Society Class III/IV)
- Left main/≥2 vessel unrevascularized coronary artery disease
- Contralateral laryngeal nerve palsy
- Severe pulmonary or renal disease

Anatomic

- Previous carotid endarterectomy with recurrent stenosis
- High cervical internal carotid artery lesions (C-2 or higher) or common carotid artery lesions below the clavicle
- Contralateral carotid occlusion
- Radiation therapy to neck
- Prior radical neck surgery
- Severe tandem lesions
- Tracheostomy

Table 5.3 High-risk features for surgical carotid revascularization.

Contraindications for carotid artery stenting

Neurological

- Major functional impairment
- Significant cognitive impairment
- Major stroke within 4 weeks

Anatomical

- Inability to achieve safe vascular access
- Severe tortuosity of aortic arch
- Severe tortuosity of common carotid artery or internal carotid artery
- Intracranial aneurysm or arteriovenous malformation requiring treatment
- Heavy lesion calcification
- Visible thrombus
- Total occlusion
- Long subtotal occlusion ("string" sign)

Clinical

- Life expectancy <5 years
- Contraindication to aspirin or thienopyridine
- Renal dysfunction

Table 5.4 Contraindications for carotid artery stenting.

The Centers for Medicare and Medicaid Services (CMS) in the United States has determined that reimbursement should be limited to qualified institutions and physicians when using FDA-approved stents and embolic protection devices (EPDs) for high-risk patients with symptomatic stenosis greater than 70%, and for high-risk patients enrolled in a Category B Investigational Device Exemption (IDE) trial or post-approval study (symptomatic stenosis >50%, asymptomatic stenosis >80%). The criteria for carotid endarterectomy (CEA) are somewhat less stringent. Symptomatic patients require more than 50% stenosis with a perioperative risk of stroke or death of less than 6%. Asymptomatic patients require more than 60% stenosis with a perioperative risk of stroke or death of less than 3% (Table 5.5).

Carotid duplex

Each ultrasound laboratory will develop its own criteria for assessing carotid artery stenosis. For example, using the Cleveland Clinic vascular

Criteria for carotid artery stenting and endarterectomy

Clinical status	CAS	CEA
Asymptomatic	>80%, high-risk, and registry	>60% and perioperative risk of stroke or death <3%
Symptomatic	>70% >50% and registry	>50% and perioperative risk of stroke or death <6%

Table 5.5 Criteria for carotid artery stenting (CAS) and endarterectomy (CEA).

Consensus criteria for noninvasive assessment of carotid artery stenosis

Stenosis	Ultrasound criteria
Normal or <50% stenosis	PSV <125 cm/sec EDV <40 cm/sec ICA/CCA PSV ratio <2 No visible plaque or intimal thickening denotes <50% stenosis
50–69%	PSV 125–230 cm/sec EDV 40–100 cm/sec ICA/CCA PSV ratio 2–4 Visible plaque
>70% or near occlusion	PSV >230 cm/sec EDV >100 cm/sec ICA/CCA PSV ratio >4 Visible plaque Markedly narrowed lumen denotes near occlusion
Total occlusion	No lumen on B-mode imaging and no flow on duplex ultrasound

Table 5.6 Consensus criteria for noninvasive assessment of carotid artery stenosis. CCA, common carotid artery; EDV, end-diastolic velocity; ICA, internal carotid artery; PSV, peak systolic velocity. Adapted from Grant EG, et al. *Radiology.* 2003;229:340-6.

medicine criteria, >80% stenosis is documented with a peak systolic velocity of >240 cm/sec and a peak diastolic velocity of >135 cm/sec (Table 5.6). The ICA to CCA ratio is also calculated from the peak systolic velocities in both vessels. A ratio ≥4 suggests a stenosis of at least 70%. There are several important conditions that the operator needs to be aware of when interpreting duplex studies.

The presence of an ICA occlusion will result in externalization of the CCA (i.e., high resistive pattern), since the flow will resemble the external carotid artery. A CCA occlusion may be reconstituted by collateral flow from the ipsilateral external carotid artery to the ICA, via the ophthalmic artery. Therefore, flow in the external artery will be retrograde and flow in the internal artery diminished. A severe carotid stenosis or occlusion may result in increased flow in the contralateral carotid artery. In this case the ratio of internal to CCA peak systolic velocity is useful. Another important collateral is the external carotid artery to the vertebral artery via the occipital branch of the external carotid. The circle of Willis can provide flow from one ICA to the contralateral hemisphere and also connect the anterior circulation (i.e., ICA) to the posterior circulation (i.e., vertebrobasilar artery). See Table 5.7 for intracranial collaterals.

Screening for CAS is indicated prior to coronary artery bypass grafting in asymptomatic patients over 65 years of age and those with left main stenosis, severe peripheral arterial disease, or prior neurological event.

Anticoagulation

Aspirin, clopidogrel, and heparin.

Carotid intervention

There are six general steps to performing carotid artery interventions:

1. Angiography of the aortic arch and carotid artery.
2. Placement of guiding catheter or guiding sheath in the CCA.
3. Preparation of all the equipment beforehand to minimize dwell time.
4. Deployment of EPD.
5. Predilation, stenting, and possibly postdilation of the lesion.
6. Retrieval of the EPD.

Prior to performing carotid intervention, characteristics that may complicate the procedure should be identified.

Intracranial collaterals

- Internal maxillary branch of the external carotid artery to the ophthalmic artery
- Facial branch of the external carotid artery to the ophthalmic artery
- Superficial temporal branch of the external carotid artery to the ophthalmic artery
- Occipital branch of the external carotid artery to the vertebral artery
- Ascending pharyngeal branch of the external carotid artery to petrous branches of the internal carotid artery

Table 5.7 Intracranial collaterals.

Difficult arch

The first area to consider is the aortic arch. Type 3 arches and bovine origin of the left common carotid are likely to need to be engaged with a Vitek or Simmon's catheter. See Figure 1.3 for the different types of arches.

Severe stenosis

Stenoses that are too tight may not allow passage of the EPD. In this case, the lesion may need to be crossed with a coronary wire and predilated with a 2.0- to 2.5-mm balloon.

Severely angulated internal carotid lesion

The problem with a severely angulated ICA lesion is that the wire may initially cross into the vessel, although the portion of the wire with the filter attached may prolapse into the external carotid artery. This can be remedied by placing a buddy wire to straighten out the system or using the Emboshield® EPD device.

Severely tortuous internal carotid lesion

May need a buddy wire (Balance Middleweight or Prowater) or even a stiff wire (Ironman or Grand Slam).

Angiography

Step 1

Aortic arch/great vessels

The aortic arch is visualized first to determine the anatomy of the great vessels. It is more difficult to engage the great vessels in type 2 and type 3 arches than type 1 arches. The arch aortogram can be performed through a variety of sheath sizes; however, a 5 Fr sheath is typically used. An angled pigtail catheter is advanced to the mid-ascending aorta and power injection performed with 15 to 20 cc/sec of contrast for a total of 30 to 40 cc in a 30° left anterior oblique (LAO) orientation.

Carotid arteries

1. Anticoagulation should be considered for angiography of the carotid arteries since the great vessels are often wired with a soft-tipped or stiff angled glide wire. An exception where anticoagulation may not be necessary would be a type 1 arch with easy to engage great vessels.

2. The CCA is usually 8–10 mm in diameter, while the ICA is 5–7 mm in diameter. The degree of stenosis of the ICA is made in reference to the nondiseased ICA segment just distal to the lesion. This is based on methodology from the North American Symptomatic Carotid Endarterectomy Trial (NASCET).

3. Selective angiography of the left carotid artery is usually performed with a JR 4 or stiff-angled glide catheter; however, a Vitek or H1 slip catheter may be needed for type 2 or 3 arches.

4. Standard views of the left carotid are taken with the patient's head turned to the right and 30° of LAO orientation.

    i. The first view can be relatively low magnification (i.e., 13") and vertically collimated so that the entire length of the carotid and the external/internal bifurcation is visualized. A 19" magnification will image from the origin of the CCA to the intracranial vessels.

    ii. The presence of any internal carotid disease is further defined with high-magnification views (i.e., 6" to 8"). The 90° lateral view is a good supplementary view; however, a straight view will sometimes be needed to separate the external carotid from the ICA.

    iii. Intracranial views are performed in a straight cranial orientation. Enough cranial angulation is given to move the eyes into the facial bones (usually about 20°) and slight left/right orientation so that the falx cerebri is visible in the mid-line.

    iv. The cranial view is useful for imaging the anterior and middle cerebral arteries and the presence of an anterior communicating artery, which fills the contralateral anterior cerebral artery. This view will also show a posterior communicating artery if present.

    v. The second intracranial view is performed with 90° lateral orientation. This is a good supplementary view for determining the presence of a posterior communicating artery.

5. Angiography of the right CCA is somewhat more difficult since it may require selective catheter engagement of the right CCA as opposed to the innominate artery. The least traumatic means of accomplishing this is to engage the innominate artery with the JR 4, or stiff-angled glide catheter, then perform "trace-subtract" in a right anterior oblique (RAO) orientation to separate the origin of the carotid artery from the subclavian artery. Figure 5.1 shows a cerebral angiogram of the right ICA, and Figure 5.2 shows the posterior circulation of the left artery.

6. A stiff-angled glide wire is advanced into the CCA, or external carotid artery if additional wire purchase is needed. With the glide wire anchored in place, the JR 4 or stiff-angled glide catheter is advanced into the CCA. This may be difficult to do if there is tortuosity of the CCA or if the patient has a type 2 or 3 arch. A wire may not be necessary if there is a straight shot to the carotid artery and the diagnostic catheter advances easily. If there is no or minimal aortic atherosclerosis, a Simmons catheter can be used, which is designed to selectively engage the common carotid; however, care must be taken with using this catheter as it is more traumatic than other catheters. A Vitek catheter can also be used, although this is shorter and may only engage the innominate artery.

7. Once the right CCA is engaged, the first view is of a relatively low magnification (i.e., 13") performed with RAO and the patient's head turned to the left. Precise anatomy of the carotid bifurcation is further defined by supplementary orientations with high magnification (i.e., 6–8"). Intracranial views are obtained as described in step 4.

Cerebral angiogram of right internal carotid artery

A

Superior division, MCA

Inferior division, MCA

M2

M1 ← A2

Terminal ICA

("T") A1

Cavernous ICA

B

Right ACA

Right MCA

Ophthalmic artery

Internal carotid artery

Figure 5.1 Cerebral angiogram of right internal carotid artery. Angiogram demonstrates segments of the right middle cerebral artery (MCA) and right anterior cerebral artery (ACA). A) anterior projection. B) lateral projection. A1 & A2, segments of the ACA; ICA, internal carotid artery; M1& M2, segments of the MCA. Reprinted with permission from Krishnaswamy A, et al. *Catheter Cardiovasc Interv.* 2010;75:530–9.

Posterior circulation

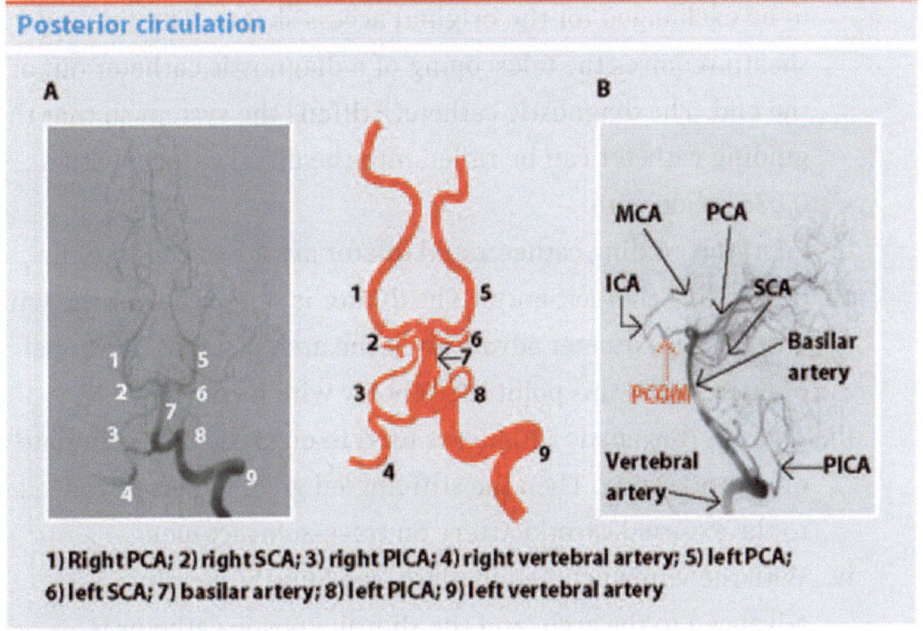

1) Right PCA; 2) right SCA; 3) right PICA; 4) right vertebral artery; 5) left PCA;
6) left SCA; 7) basilar artery; 8) left PICA; 9) left vertebral artery

Figure 5.2 Posterior circulation. A) Selective angiogram of a dominant left vertebral artery. B) Lateral projection of a vertebral artery injection with filling of the anterior circulation via the PCOM. AICA, anterior inferior cerebellar artery; ICA, internal carotid artery; MCA, middle cerebral artery; PCA, posterior cerebral artery; PICA, posterior inferior cerebellar artery; PCOM, posterior communicating artery; SCA, superior cerebellar artery. Reprinted with permission from Krishnaswamy A, et al. *Catheter Cardiovasc Interv.* 2010;75:530–9.

Vertebral arteries

The vertebral arteries are best engaged with an atraumatic catheter such as a stiff-angled glide catheter or vertebral catheter.

Step 2

1. Heparin is administered to achieve an activated clotting time (ACT) of >270 seconds.
2. There are two options for guide support: a guiding sheath or a guide catheter.
3. The preferred system is the use of a guiding sheath (i.e., Shuttle sheath), which combines the access sheath and guiding catheter on the same device.
    i.   A typical size for the guiding sheath would be 6 Fr and 80–90 cm length. The 80 cm sheath can be considered for shorter patients. The shorter length is also necessary if an MP catheter is needed for aspiration in the event of no-reflow after stent deployment. When the Shuttle is used, it needs to be exchanged for the original access sheath. The guiding sheath requires the telescoping of a diagnostic catheter out of the end. The diagnostic catheter stiffens the system so that the guiding catheter can be railed into the distal carotid with a 0.035" glide wire.
    ii. First, the guiding catheter and dilator are advanced into the descending thoracic aorta. The dilator is removed over the wire and the slip catheter advanced in the arch distal to the vessel of interest). At this point the Wholey wire is removed.
    iii. The slip/diagnostic catheter is used to engage the innominate or carotid artery. Then the stiff-angled glide wire is advanced to the external carotid artery on trace-subtract mode.
    iv. With the wire anchored in place, the Shuttle sheath is advanced to the arch, and the slip/diagnostic catheter is then advanced into the distal CCA. With the wire and slip/diagnostic catheter held stationary, the Shuttle sheath is railed to the distal common carotid, following with wire and diagnostic catheter are carefully removed. This system

provides maximal support, although it may be more difficult to position the guiding sheath with a type 3 arch.

v. To position the guiding sheath atraumatically in patients with type 3 arches, a hydrophilic slip catheter (H1 or Vitek catheter) is telescoped out of the end of the guiding catheter. A long JR 4 catheter can also be used; however, it is not hydrophilic. The slip catheter is used to engage the artery of interest. A stiff-angled 0.035" glide wire is advanced into the external carotid artery on trace-subtract mode, which is then anchored. This allows the slip/diagnostic catheter to rail into the CCA. The guiding sheath is then advanced to the distal carotid artery over the slip catheter.

4. The alternative option is to use a guide catheter. This system is easier to use, although support is less than maximal. Since this catheter is rigid, it is generally not intended to be railed into the common carotid. The guide catheter is best if a Shuttle sheath is unable to be advanced into the carotid artery due to an extreme type 3 arch. The typical example would be an 8 Fr H1 guide catheter that is used to engage the proximal innominate or left CCA.

Step 3

The goal of this step is to reduce dwell time, since increased dwell times are associated with adverse outcomes.

Equipment
1. Predilation balloon (e.g., 4 x 20 mm Voyager™).
2. Postdilation balloon (e.g., 5 x 20 mm Aviator or Viatrac).
3. EPD.
4. Stent (types of carotid stents are listed in Table 5.8).
5. EPD retrieval sheath.

It is extremely important that the EPD is de-aired and the stent is flushed prior to use.

Types of carotid stents

Stent type	Manufacturer	Name	Tapered stent proximal/distal diameter (mm); length (mm)	Straight stent diameter (mm); length (mm)
Stainless steel	Boston Scientific	Wallstent®*	NA	6 (X 22), 8 (X 21, 29, 36), 10 (X 24, 31, 37)
Open-cell nitinol	Guidant	Acculink*	10/7, 8/6; 30, 40	5, 6, 7, 8, 9, 10; 20, 30, 40
	Medtronic	Exponent®*	20, 30, 40	6, 7, 8, 9, 10; 20, 30, 40
	Bard	Vivexx™	12/8, 10/7, 8/6; 30, 40	5, 6, 7, 8, 9, 10, 12; 20, 30, 40
	ev3	Protégé*	10/7, 8/6; 30, 40	6, 7, 8, 9, 10; 20, 30, 40, 60
	Cordis	Precise*	NA	5, 6, 7, 8, 9, 10; 20, 30, 40
Closed-cell nitinol	Endotex	NexStent™†	NA	4, 5, 6, 7, 8, 9; 30
	Abbott Vascular	Xact*	10/8, 9/7, 8/6; 30, 40	7, 8, 9, 10; 20, 30
	Medinol	Nirtinol	10/7, 8/6; 30, 44	5, 6, 7, 8; 21, 30, 44

Table 5.8 Types of carotid stents. *US Food and Drug Administration approved as of January 2010. †US Food and Drug Administration class I recall in June 2008.

Step 4

Road-mapping (also referred to as trace-subtract) is performed in an oblique orientation to delineate the carotid bifurcation. Typically two curves are placed in the distal tip of the wire, unlike coronary intervention. The EPD (on a 0.014" wire) is advanced to the mid ICA and the EPD is deployed. This begins the dwell time.

Step 5

Predilation of the lesion serves as a good gauge of the degree of hemodynamic instability that may occur with stenting and postdilation. Marked bradycardia or asystole with predilation may require atropine for stenting and postdilation. Inserting the stent into the sheath can entrain air, which is remedied by frequent bleed-backs from the catheter after all device exchanges. The stent is positioned across the lesion and carefully deployed. The self-expandable stent has a tendency to come back with deployment, which can be remedied by initiating deployment slightly more distal to the lesion.

Step 6

The last step is retrieval of the EPD with the provided retrieval sheath. This ends the dwell time.

See Table 5.9 for potential complications and their incidence following CAS. Patients who undergo CAS are at risk of hyperperfusion due to impaired cerebral autoregulation. Patients with hyperperfusion may have a severe headache, marked hypertension, and often nausea, vomiting, and/or seizures. The risk factors for hyperfusion are shown in Table 5.10.

Pointers

- Difficult interventions include type 2 or 3 arches. The 6 x 90 cm Shuttle sheath may not provide enough support to allow passage of a device to the lesion. This can be remedied by placing a 0.014" buddy wire in the ICA, or by up-sizing to an 8 Fr H1 guide catheter for maximal support.
- Occasionally, the retrieval sheath may not advance to the EPD to allow retrieval. When this occurs a 5 Fr MP catheter can be used as the retrieval device.
- Severe angulation of the ICA may not allow the EPD to advance. For example, the soft portion of the wire may engage the carotid; however, the stiffer portion of the wire with the filter may prolapse into the external carotid. A 300-cm 0.014" buddy wire should be placed in the distal internal carotid to straighten the common/internal carotid arteries enough to allow the EPD to pass. If this is not sufficient, a 1.5-mm over-the-wire balloon or exchange catheter (e.g., Excelsior) can be advanced to the distal ICA, and the original wire exchanged for a Grand Slam wire. This wire has a soft tip, although it is very rigid and should straighten the common/internal carotid arteries to allow the EPD to pass.

Potential complications and their incidence following carotid artery stenting

Cardiovascular	
Vasovagal reaction	(5–10%)
Vasodepressor reaction	(5–10%)
Myocardial infarction	(1%)
Carotid artery	
Dissection	(<1%)
Thrombosis	(<1%)
Perforation	(<1%)
External carotid artery stenosis or occlusion	(5–10%)
Transient vasospasm	(10–15%)
Restenosis	(3–5%)
Neurological	
Transient ischemic attack	(1–2%)
Stroke	(2–3%)
Intracranial hemorrhage	(<1%)
Hyperperfusion syndrome	(<1%)
Seizures	(<1%)
General	
Access site injury	(5%)
Blood transfusion	(2–3%)
Contrast nephropathy	(2%)
Contrast reactions	(1%)
Death	(1%)

Table 5.9 Potential complications and their incidence following carotid artery stenting.

Risk factors for hyperperfusion

- Critical (i.e., >90%) internal carotid artery stenosis
- String sign of the internal carotid artery
- Severe bilateral lesions
- Isolated hemisphere due to poor collateral circulation
- Uncontrolled hypertension

Table 5.10 Risk factors for hyperperfusion.

Examples of filter-type embolic protection devices used for carotid intervention

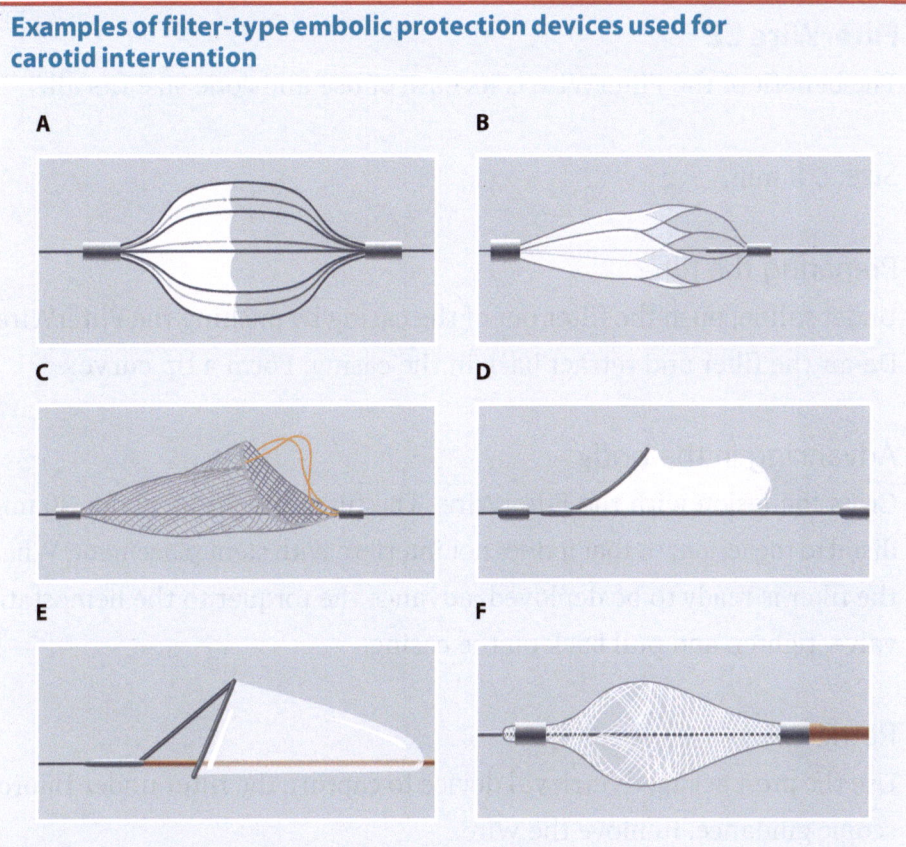

Figure 5.3 Examples of filter-type embolic protection devices used for carotid intervention.
A) AngioGuard® XP (Cordis); B) Accunet™ (Guidant Corporation); C) Spider™ (ev3); D) FilterWire™ Ex (Boston Scientific); E) FilterWire EZ (Boston Scientific); F) Interceptor® (Medtronic).

Embolic protection devices

Embolic protection is indicated for saphenous vein graft interventions where it has been demonstrated to reduce major adverse cardiac events. Embolic protection is also routinely used in carotid intervention and increasingly in lower extremity revascularization despite a lack of available randomized clinical data demonstrating its use. It is important to have a good working knowledge of the various EPDs, as different devices are suitable in different circumstances.

See Figure 5.3 for examples of filter-type EPDs used for carotid intervention. Table 5.11 shows a comparison of selected distal EPDs.

FilterWire EZ

The benefit of the FilterWire is its ease of use and "one size fits all."

Size: 5.5 mm.

Preparing the filter

Under saline, push the filter out of the casing by pushing the FilterWire. De-air the filter and retract back in the casing. Form a tip curve.

Advancing in the body

Cross the lesion with the FilterWire. The filter should be about 30 mm distal to the lesion, so that it does not interfere with stent placement. When the filter is ready to be deployed, advance the torquer to the hemostatic valve, tighten and pull back on the casing.

Retrieval

Use the pre-packaged retrieval device to capture the filter under fluoroscopic guidance. Remove the wire.

Comparison of selected distal embolic protection devices

	Spider	FilterWire	AngioGuard*†	Accunet*
Manufacturer	ev3	Boston Scientific	Cordis/ J&J, Inc.	Guidant
Material	N	N, PU	N, PU	N, PU
Guidewire (inches)	0.018	0.014	0.014	0.014
RX	Yes	Yes	Yes	Yes
Independent wire	Yes	No	No	No
Vessel size (mm)	3–6	3.5–5.5	4–8	4.5–7.5
Profile (French)	3.2	3.2	3.2–3.9	3.5–3.7
Pore size (microns)	167–209	110	100	150

Table 5.11 Comparison of selected distal embolic protection devices. *Food and Drug Administration approved as of November 2006. †Embotrac guidewire is an independent wire that must be used with the Emboshield filter to prevent filter migration and embolization. ‡Rubicon and Interceptor are deployed by remote activation, rather than by retraction of a distal constraining sheath. EPD, embolic protection devices; FEP, fluorinated ethylene propylene; PTFE, polytetrafluoroethylene; N, nitinol; PU, polyurethane.

Spider RX

Sizes: 3, 4, 5, 6, and 7 mm.

Preparing the filter

The Spider RX EPD casing has both a green and a blue end. Focus on the green end first. Push the filter out of the green end of the casing under saline. Making sure there are no bubbles, pull the filter into the casing, all the way to the proximal end of the clear portion.

Advancing in the body

Cross the lesion with a long standard workhorse wire. Advance the green end of the Spider on the workhorse wire and exit this wire out of the side port of the casing. The exit port of the Spider monorail is at the transition of the green and clear portion of the casing on the more distal part of the casing. There is a radiopaque marker at the distal tip of the green end of the casing. Advance the casing 4 cm distal to the lesion. At this point, remove the workhorse wire and advance the filter (two radiopaque markers) to the end of the casing. The distal filter marker should be lined up with the casing marker. When the filter is ready to

Emboshield*	Interceptor‡	Gore Filter System (GFS)	Rubicon‡
Abbott Medical	Medtronic	Gore	Boston Scientific
N, PU	N	N, PTFE, FEP	N, PU
0.018	0.014	0.014	0.014
Yes	Yes	Yes	No
Yes	No	No	No
3–6	4.5–6.5	2.5–5.5	3–6
3.9	2.7	3.2	2.1–2.7
140	100	100	100

be deployed, unsheath it by keeping the wire pinned and retracting the casing. Remove the casing from the body. Treat the lesion.

Retrieval of the filter
Use the blue end of the casing to retrieve the filter. The gold thread that is embedded in the filter should be pulled into the casing. Remove the wire and retrieval device as a unit.

Acculink
Sizes: 5.5, 6.5, and 7.5 mm.

Preparing the filter
Hold the clear plastic housing, which surrounds the filter in an upright position. Inject saline until it comes out of the top. Tighten the screw on the top of the housing until saline fills the opaque plastic coil. Pull the wire so that the filter completely enters into the light blue microsheath. At this point loosen the screw and remove the filter from the plastic housing. Confirm that there is no air in the filter by opening under saline. One person holds the tip under saline, while a second person at the back of the wire pushes the filter out of the light blue microsheath. Pull the filter back into the microsheath. Tighten the Acculink torquer over the light blue microsheath on the back of the wire.

Wire
Wire the lesion in the usual fashion.

Retrieval
Always use the retrieval 2 device. This is softer than the retrieval 1 device.

AngioGuard
Sizes: 4, 5, 6, 7, and 8 mm.

Preparing the filter
The filter system comes wrapped in a plastic coil. Hold the clear plastic portion that contains the filter upright. Inject saline through the clear

plastic portion and then tighten the valve. Release the two antimigration clips closest to the filter, tighten the torquer, and pull the wire until the filter is completely inside the casing. On the over-the-wire system, the torquer must be advanced further to the green microsheath until it can be tightened. Tightening the torquer on the back of the microsheath prevents the casing from splitting. Release the remaining antimigration clip, loosen the valve on the clear plastic filter container and pull the wire back so that the whole system comes out of the plastic coil. Form a tip on the wire, being careful not to unsheath the filter.

Wire

Cross the lesion with the wire. The wire introducer that comes with this system is very bulky and will not fit in the hemostatic valve if a buddy wire has been used. In this case, use a separately packaged plastic wire introducer if available in the lab. When the filter is ready to be deployed in the pre-petrous region of the internal carotid, straighten the wire and walk the casing backward over the wire. Split the black casing to the body all the way to the clear portion, being careful to leave the filter in place (i.e., watch under fluoroscopy). Completely remove the black casing.

Retrieval

Advance the retrieval device to the filter and capture the filter. Remove the wire and filter.

Emboshield

This EPD is primarily used for carotid interventions. The lesion is crossed with a BareWire. This wire has a 0.019" step on the distal end, which prevents filter embolization. This is distinct from other filters (except Spider RX) where the filter is attached to the wire. This is advantageous with a severely angulated or tortuous ICA. A potential disadvantage of this system is that if the wire gets pulled back significantly, the filter will also get pulled back; however, unlike other filters the Emboshield cannot be re-advanced just by re-advancing the wire. Where this EPD is used in the setting of a research study, it requires the use of the Xact stent, which is quite rigid. This stent delivery system has a maximal outer diameter of 5.7 Fr.

Sizes: 3, 4, 5, and 6 mm.

Preparing the filter

Fill the plastic wells with saline. Use the syringe to inject saline inside the filter and then squeeze the filter to remove any residual air. Pull the filter into the casing. This will cause the wire to bow out from the spool. Remove the device. Make sure the filter is completely covered.

Advancing in the body

The Emboshield BareWire (soft, medium, firm, super, and maximum), which is 0.014" with a 0.019" step 30 mm proximal to the tip, must be used. Cross the lesion in the ICA to the pre-petrous portion. Advance the Emboshield over the wire close to the distal wire marker (i.e., 10 mm). The filter should also be 15 mm from the intended distal margin of the stent. This ensures that if the wire position changes slightly, the filter will remain in place. The filter has two markers, so there will be a total of three markers when the filter is on the wire. When the filter is in the correct location, remove the red safety cap on the back of the filter device, keep the device stationary, pull the white tab back to unsheath the filter, then walk the delivery device back. Be aware that the filter is loose on the wire and should therefore be relatively fixed in place by contact with the vessel wall. If the filter migrates proximally it can not be easily re-advanced. If it is critical to advance the filter forward, this can be done with the retrieval sheath without removing the red cap.

Retrieval

Retrieval is a two step process. The first step retracts the centering catheter, while the second step brings the filter into the expansile pod of the retrieval catheter. First, advance the retrieval device (3.7 Fr) proximal to the filter. Gently pull the wire so that the 0.019" step makes contact with the filter. Remove the red safety cap and pull the black tab back. This retracts the centering catheter. A marker proximal to the expansile tip will become visible. Next, pull the BareWire with the wire torque device attached, so that the filter enters into the expansile tip. The proximal marker of the filter should line up with the marker on the retrieval catheter. Remove the wire and retrieval device as a unit.

Other considerations

A screw-type hemostatic valve, such as O-ring, is preferable to a Copilot hemostatic valve. The former allows more generous bleeding and prevents air embolism. Additionally, when the stent is ready to be deployed, tighten the hemostatic valve down on the stent delivery catheter. This prevents movement of the stent during deployment. The Emboshield EPD is intended to be used with the Xact stent (crossing profile of 6.7 Fr).

Iliac artery interventions
Indication

Lifestyle limiting disability due to intermittent claudication in patients who have (a) an inadequate response to exercise or pharmacological

TASC morphological stratification of Iliac lesions	
Type A	Single stenosis of the common or external iliac artery <3 cm long (unilateral or bilateral)
Type B	Single stenosis of the common or external iliac artery 3–10 cm long, but not extending into the common femoral artery
	Two stenoses in the common or external iliac artery <5 cm long, but not involving the common femoral artery
	Unilateral common iliac artery occlusion
Type C	Bilateral stenosis of common and/or external iliac artery 5–10 cm long, but not involving the common femoral artery
	Unilateral external iliac artery occlusion, not involving the common femoral artery
	Unilateral external iliac artery stenosis, that extends into the common femoral artery
	Bilateral common iliac artery occlusion
Type D	Diffuse stenosis of the entire common, external iliac and common femoral artery >10 cm long
	Unilateral occlusion of the common and external iliac arteries
	Bilateral external iliac artery occlusion
	Iliac stenosis adjacent to aortic or iliac aneurysm

Table 5.12 TASC morphological stratification of Iliac lesions. TASC, Trans-Atlantic Inter-Society Consensus. Dormandy JA, et al. *J Vasc Surg.* 2000;31:S1-S296.

therapy, and/or (b) a favorable riskbenefit ratio (e.g., focal aortoiliac occlusive disease).

A percutaneous approach is the preferred technique for Transatlantic Inter-Society Consensus (TASC) Working Group type A iliac and femoropopliteal arterial lesions. There is no consensus regarding the optimal management of type B and C lesions. Type D lesions should be surgically revascularized (Table 5.12).

Anticoagulation

Aspirin, clopidogrel, and heparin.

Angiography

The distal abdominal aorta and iliac arteries are visualized nonselectively by distal abdominal aortogram. This can be performed through a variety of sheath sizes; however, a 5 Fr sheath is most typical. A straight flush catheter or straight pigtail catheter is advanced over a soft-tip 0.035" wire to the distal abdominal aorta (T12 to L1 level) and power injection performed with 15–20 cc/sec of contrast for 30–40 cc. The distal iliac and femoral artery bifurcation can be visualized non-selectively by pulling the catheter closer to the iliac bifurcation and power injection performed with 7–10 cc/sec of contrast for 14–20 cc. The iliac bifurcation is best seen with contralateral angulation, while the femoral bifurcation is best seen with ipsilateral angulation.

Intervention

1. Translesional pressure gradients (with and without vasodilation) should be obtained to evaluate the significance of angiographic iliac arterial stenoses of 50–75% diameter before intervention.
2. Revascularization of the iliac artery usually takes place in a retrograde fashion from the ipsilateral groin; therefore, bilateral iliac lesions will require bilateral sheaths.
3. A 35-cm 7 Fr sheath (e.g., Brite Tip) is inserted into the common femoral artery, proximal to the iliac lesion. This leaves part of the sheath externalized. Full insertion of a 35 cm sheath places the tip in the abdominal aorta and distal to the lesion.

Indications for stenting in aorto-iliac disease
• Provisional stenting for suboptimal results of percutaneous angioplasty (either from extensive dissection, residual diameter stenosis >50%, or residual translesional gradient ≥10 mm Hg) (Class I)
• Total occlusion of common or external iliac arteries (Class I)
• Can be considered for all aorto-iliac lesions (Class I)
• Recurrence after prior angioplasty
• Ostial location of lesion
• Severe calcification

Table 5.13 Indications for stenting in aorto-iliac disease.

4. A soft-tip wire or stiff-angled glide wire is advanced across the lesion into the descending abdominal aorta.

5. Predilation of the lesion can be performed with a 6-mm diameter balloon, stented according to anatomy, and post-dilated if necessary.

6. If a self-expandable stent is used, this requires placement over a 0.018" wire. Self-expanding stents should be considered in all nonostial common iliac and all external iliac artery lesions.

See Table 5.13 for indications for stenting in aortoiliac disease.

Superficial femoral artery intervention

Indications

1. Lifestyle limiting disability due to intermittent claudication, in patients with an inadequate response to exercise or pharmacological therapy.

2. Ischemic tissue loss.

Anticoagulation

Aspirin, clopidogrel, and heparin.

Angiography

1. The superficial femoral artery and profunda artery are visualized by distal abdominal aortogram with lower extremity run-offs or selective angiography of the common iliac with static run-offs. If noninvasive testing reveals bilateral lower extremity disease, the lower extremity run-off is a good starting point. This can be

performed through a variety of sheath sizes; however, a 5 Fr sheath is most typical. A straight flush catheter or straight pigtail catheter is advanced over a soft-tip 0.035" wire to the distal abdominal aorta (T12 to L1 level) and power injection performed with 15 cc/sec of contrast for 90 cc.

2. If a lesion is known or suspected on one leg, selective angiography of the ipsilateral common iliac artery with static run-offs of the superficial femoral artery can be performed from the contralateral groin. For example, where noninvasive testing reveals severe right-sided thigh disease, a 5 Fr sheath is best placed in the left common femoral artery, and a 5 Fr internal mammary artery catheter engaged in the right common iliac artery (over a soft-tipped 0.035" wire or angled glide wire).

3. Digital subtraction angiography on low magnification (i.e., 17–19") then takes place on the right superficial femoral artery with 10-cc hand injection (diluted by half with normal saline).

4. A metal clamp or hemostat can be placed on the inferior margin of this image, which will serve as a reference for the superior margin of the thigh vessels.

5. For imaging the distal superficial femoral artery, it is best to exchange the internal mammary artery catheter for a straight flush catheter and advance into the superficial femoral artery. This is less likely to cause dissection with injection and there is also less waste of contrast into the profunda artery.

6. Severe iliac and/or femoral artery lesions that prevent accessing the common femoral artery may require arm access. It is easier to advance a catheter to the distal abdominal aorta from the left brachial artery compared to the right brachial artery. From the brachial artery, a 100 cm angled catheter will reach the ostium of the common iliac arteries, while a catheter from the radial artery will reach the mid-descending abdominal aorta.

Intervention

1. Revascularization of the proximal superficial femoral artery is performed in an antegrade fashion from the contralateral groin.

A soft-tipped 0.035" wire is advanced from the internal mammary artery catheter (engaged in the involved common iliac artery) to the level of the common femoral artery. Occasionally, a stiff-angled glide wire is required due to vessel tortuosity and calcification. Use of this wire carrries a higher risk of dissection.

2. The wire is left in place, the diagnostic catheter removed, and a 6–7 Fr sheath inserted to the level of the common femoral artery. The Brite Tip sheath is preferable but the Balkin sheath can be used when maximal support is needed.

3. For minimally splayed iliac arteries, the soft-tipped wire may not provide enough support to rail the sheath. In this case, an extra-stiff Amplatz (1-cm tip) wire can be used. However, the increased utility of this wire comes at the cost of increased vessel trauma.

4. For distal superficial femoral artery lesions, an MP diagnostic catheter can be telescoped out of the end of the sheath just proximal to the lesion. This will minimize damage to the nondiseased superficial femoral artery from multiple exchanges of balloon and stents. Dilation of the lesion can be performed with a long 4-mm diameter balloon. Stenting should be considered when results after percutaneous angioplasty are suboptimal, either from extensive dissection, residual diameter stenosis >50%, or residual translesional gradient ≥10 mmHg.

Specialized devices for peripheral vascular interventions

Directional atherectomy: SilverHawk®

The SilverHawk is not indicated for heavily calcified lesions. It can be used for in-stent restenosis, although this should be used with caution due to the risk of it getting caught in a stent strut. The device is most appropriate for softer, atherosclerotic lesions. Directional atherectomy often produces a standalone result; therefore, it can be used to avoid stent placement, which makes it an attractive option across joint spaces.

The SilverHawk uses a 0.014" wire and comes in a variety of sizes. "L" (large vessels; 5–7 mm) is the standard size for the superficial femoral artery; and "M" (4 or 5 mm) or "S" (<4 mm) can be used for popliteal and infra-popliteal arteries, respectively.

Intervention

1. Advance the cutter over the wire just proximal to the lesion in the off position. When atherectomy is ready to be performed, place the device in the on position and pull the thumb trigger to activate the cutter. Advance across the lesion up to 5 cm for each pass.

2. Push the thumb trigger forward in the neutral position (this also packs the nose cone). The packer marker will move closer to the cutter as the nose-cone is filled. Pull the device back to the starting point.

3. Rotate the cutter with the torquer on the hand unit to prepare for the next pass. One click rotates the tip 10°; therefore, turn the torquer 9 clicks, or 90°, for each quadrant, when the packer is in 4 quadrants (90°). If the nose-cone is not full, additional passes can be made prior to removing the device from the body.

4. The device should be turned off and in the neutral position to remove it from the body. The device can be left on the wire to clean it, although it is easier to remove it from the wire.

5. Once the device is removed from the wire:
   i. Turn the device back on.
   ii. Pull the thumb trigger so that the cutter is activated.
   iii. Turn the device off so that the cutter port remains open.
   iv. Insert the included cleaner needle to the nose cone and flush the atherothrombotic debris out of the cutter port.

Orbital atherectomy: Diamondback®

The orbital atherectomy is very similar to rotational atherectomy. The notable difference is that the "burr" for rotational atherectomy spins with a tight axis of rotation that is centered over the wire, while the "crown" for orbital atherectomy spins with an orbital axis so that it conforms to the inner circumference of the vessel wall, even if oblique or eccentric. The result of orbital atherectomy is that a larger effective vessel size can be atherectomized. Rotational atherectomy is used for the coronaries, while orbital atherectomy is well-suited for the vessels of the lower extremities.

Crown and estimated lumen size with Diamondback

Crown size (mm)	80,000	140,000	200,000
Classic crown			
1.25	1.5	1.8	2.1
1.50	1.8	2.2	2.6
1.75	2.0	2.4	2.9
2.00	2.2	2.9	3.4
2.25	2.5	3.3	3.8
Solid crown			
1.50	1.65	2.10	2.70
1.75	2.05	2.70	3.45
2.00	2.65	3.45	4.00
2.25	2.75	3.65	4.50

Table 5.14 Crown and estimated lumen size with Diamondback. Estimated lumen size is based on 1 minute of sanding for a given crown size and speed of rotation.

The typical vessel for orbital atherectomy is the superficial femoral artery, which is frequently heavily calcified. The Diamondback wire has to be delivered secondarily.

Intervention

1. The general setup is to cross the lesion with a workhorse 0.014" wire and exchange this wire through a 4 Fr stiff-angled glide catheter for the Diamondback wire. This wire is 0.014"; however it has a thicker tip (0.22") and it cannot accommodate a curve. Given the thick tip on this wire, it is not possible to deliver it through a microcatheter.
2. Select a crown size (Table 5.14), advance it just proximal to the lesion and begin sanding the area of interest. Each pass should be limited to 1 minute (Table 5.14).

Peripheral re-entry devices
Outback®

Intervention of a superficial femoral artery chronic total occlusion often results in subintimal placement of the wire. When this happens, a re-entry device can be used to re-establish wire placement in the true lumen.

Intervention

1. The false lumen needs to be crossed with a stiff 0.014" wire distal to the intended re-entry site. Since re-entry devices are bulky, the subintimal space may need to be dilated with a 2- or 3-mm balloon to facilitate passage.

2. The Outback device is prepared on the back table and advanced over-the-wire to the intended re-entry site. Correct placement of the device is confirmed in orthogonal views. Use 45° LAO and RAO projections.

3. In one projection, the tip of the Outback should look like a "T" over the artery during cine, while in another projection, the tip should look like a "L." The "L" points in the direction where the needle will exit.

4. The correct position is confirmed, the wire is withdrawn proximal to the needle, the needle is deployed, the true lumen is wired, and the device is removed from the body.

Pioneer®

The Pioneer is similar to the Outback in that a needle exits from the false lumen to the true lumen, although IVUS is used to establish correct device placement.

Intervention

1. The false lumen is wired distal to the intended re-entry site, the Pioneer is advanced by rapid exchange into position and the IVUS activated. The true and false lumen should be visualized.

2. Rotate the device so that the true lumen is at "12 o-clock." This is where the needle will exit.

3. Advance a second 300-cm wire through the back of the device ("over-the-wire") proximal to the needle.

4. The needle can exit at 3, 5, or 7 mm. Start with 3 mm, deploy the needle, and attempt to enter the true lumen with the "over-the-wire."

5. If unsuccessful, retract the needle and re-attempt at 5 mm and 7 mm if needed.

Subclavian artery intervention

Indication

1. Subclavian artery stenosis with arm claudication or ischemia.
2. Asymptomatic subclavian artery stenosis with a gradient of at least 20 mmHg.
3. Subclavian steal from vertebral or left internal mammary artery graft.

Percutaneous intervention is contraindicated in the presence of fresh thrombus.

Angiography

The great vessels are engaged using an LAO orientation, although a projection contralateral to the vessel will "spread-out" the proximal portion of the subclavian artery.

Access

Right or left common femoral artery.

Anticoagulation

Aspirin, clopidogrel, and heparin.

Equipment

Guiding sheath technique

1. 6/80-cm guiding sheath with hemostatic valve.
2. 5 Fr 125-cm diagnostic catheter (JR 4, stiff-angled glide catheter).
3. 300-cm soft-tipped 0.035" wire.

Guide catheter technique

1. 7 Fr 100-cm guide catheter with hemostatic valve (JR, MP).
2. 300-cm soft-tipped 0.035" wire.

Intervention

For the guiding sheath technique begin with a short 5 Fr sheath in the common femoral artery. Follow steps 1–5 below.

For the guiding catheter technique begin with a short 7 Fr sheath in the common femoral artery. Use a 300 cm soft-tipped 0.035" wire to advance a 7 Fr MP or JR catheter past the origin of the great vessel. Remove the wire and engage the vessel with the catheter. Follow steps 3–5 below.

1. Prepare the guiding sheath. The Shuttle sheath in particular comes packed as two parts: the Shuttle sheath and the dilator. Flush both. Attach a 3-way stopcock to the hemostatic valve, which is then attached to the Shuttle sheath. Insert the dilator through the hemostatic valve. The dilator will not snap in place; therefore, hold it firmly while advancing in the body.

2. Advance a 300-cm soft-tipped 0.035" wire to the proximal thoracic aorta and remove the short 5 Fr sheath. Advance the sheath/dilator over the wire so that the tip of the Shuttle is in the proximal thoracic aorta. Leave the wire in place, remove the dilator, and de-air the Shuttle.

3. Push the wire to the ascending aorta and advance the diagnostic catheter through the guiding sheath and past the great vessel of interest. Remove the wire and connect the diagnostic catheter to pressure. Engage the innominant or left subclavian artery with the diagnostic catheter. Advance the soft-tipped 0.035" wire across the lesion to the axillary artery. A severe subclavian artery stenosis may be difficult to cross with the 0.035" wire. In this case a 0.035" stiff-angled glide wire or a 0.014" Balance Heavyweight wire can be used.

4. With the wire as a rail, the diagnostic catheter is advanced across the lesion into the subclavian artery and the Shuttle sheath is advanced to the ostium of the subclavian. The diagnostic catheter is now removed and the wire left in place. If a glide wire or 0.014" wire was required to cross the lesion, the diagnostic catheter is advanced across the lesion and the wire exchanged for the soft-tipped 0.035" wire. At this point, the diagnostic catheter is removed.

5. Pre-dilate the lesion (i.e., 6 x 20 mm balloon). Stent the lesion (i.e., 7 x 20 mm or 8 x 20 mm [or longer if necessary]). Examples include Genesis, Palmaz Blue stent, and OmniLink.

Chapter 6

Structural heart disease

Aortic stenosis

Diagnosis

A quality transthoracic echocardiographic study produces excellent and reproducible determination of the degree of aortic stenosis, and therefore, in most cases obviates the need for invasive hemodynamic study. However, in cases where transthoracic echo is not as reliable, such as morbid obesity with poor image quality, the catheterization laboratory may be important for determining the degree of stenosis. A simple pull-back across the aortic valve is often performed to yield a peak-to-peak gradient; however, this is just a crude estimate of the mean gradient. Preferentially, dual transducers should be employed to accurately determine the mean gradient.

Anticoagulation

While anticoagulation is not routinely employed, a severely calcified aortic valve carries a potential stroke risk upon crossing it. Therefore, consideration should be given to heparinization prior to study for suspected severe aortic stenosis.

Procedure

1. Passage across the valve with a 5 or 6 Fr pigtail catheter and a regular J-wire can initially be attempted. The inability of the pigtail catheter to cross the valve is an immediate sign that severe

aortic stenosis may be present. In this case, an AL 1 or 2 catheter and a straight tip 0.035" wire are often used.

2. With the camera in the LAO projection, the catheter is gently clocked or counter-clocked, and the straight wire is gently advanced until it enters into the left ventricle (LV). Gentle bobbing movements of the catheter as it enters the jet stream of flow is a sign that it is nearing the aortic orifice.

3. Each attempt at crossing the valve should not last longer than 2 or 3 minutes, at which time the catheter is aspirated and the blood discarded. Extreme care should be taken not to advance the wire down the right coronary artery or the left main trunk.

4. Once the wire advances into the LV, the catheter is advanced into the mid-ventricle and the pressure transduced.

   The following systems can be used to obtain simultaneous LV and aortic (Ao) pressures:

   i. **One arterial sheath.** A dual-lumen pigtail can be placed in the LV. Following the basic procedure above, a long exchange wire is placed through the catheter that was used to cross the valve and the catheter is removed from the body. Over this long exchange wire, a pigtail catheter is advanced into the LV and the wire removed.

   If a dual-lumen pigtail is not available, an alternative is to use a sheath that is one or two sizes larger than the catheter. This way, a standard pigtail can be advanced into the LV, while the sheath pressure is transduced. This may take some experimentation to determine what combination of equipment should be used. For example, a 4 Fr pigtail catheter works well with a 5 Fr sheath, although subtleties in different equipment brands may make a difference.

   ii. **Two arterial sheaths.** The pigtail catheter is advanced from one sheath, while the arterial pressure is transduced from the other sheath. The downside of this approach is that it requires two arterial sticks. Note that the peak femoral pressure is often higher than the central Ao pressure due to reflected pressure along the aorta. This tends to underestimate the pressure gradient across the aortic valve. Using a long sheath helps to correct this.

The Hakki formula is used to determine the aortic valve area:

aortic valve area in cm² = cardiac output (L/min)/(square root of the mean gradient)

Cardiac output requires pulmonary artery saturation and therefore right heart catheterization.

Aortic valvuloplasty

Indication

Aortic valvuloplasty is considered a Class IIb indication, as a bridge to aortic valve replacement (AVR) or as a palliative procedure. It can also be used as a bridge to AVR or percutaneous AVR to test if relieving aortic stenosis improves the patient's health and/or quality of life and to assess if it is worth proceeding with AVR or percutaneous AVR with the increased associated risks. Aortic valvuloplasty can help guide a decision about further therapies or another balloon aortic valvuloplasty procedure.

Anticoagulation

Heparin.

Arterial access

Right femoral artery (above bifurcation to allow for 12 Fr sheath). We usually insert a 4 Fr (micropuncture) sheath first, and perform angiography to confirm that the stick is above the bifurcation. If it is low, the 4 Fr sheath is removed, and manual hemostasis is achieved. Femoral access should then be reattempted. Before upsizing to a 12 Fr sheath, perform coronary angiogram if needed, and if there is question of the severity of aortic stenosis, measure gradients. You can also upsize to a 8 Fr sheath prior to investigation, and then either perclose on the way to upsizing to 12 Fr or place a single perclose if there is no balloon aortic valvuloplasty.

Intervention

1. Perclose:

    i. Make an incision with a blade and spread with Kelly clamp before inserting the first Perclose Proglide over the 0.035" wire. Insert the perclose and go through the steps as if attaining hemostasis up to the point where the strings are removed.

    ii. At this point, remove the strings and attach with the Kelly clamp to the towel.

    iii. Reinsert the wire and remove the first perclose device.

    iv. Insert a second perclose device rotated 90° from the first perclose alignment and repeat the steps above.

    v. Insert a 12 Fr Cook short sheath (using two dilators; an 8 Fr sheath is the same size as a 10 Fr dilator only). A long sheath may be needed if the patient is obese or if there is peripheral vascular disease. For a 25-mm balloon, a larger sheath (14 Fr) may be necessary. If the 12 Fr sheath (especially if it is not Cook) is bleeding, insert an 8 Fr sheath and dilator partially through it to stop the bleeding.

2. Using right femoral venous access, float a Swan-Ganz catheter into the pulmonary artery. The S-shape curve of this catheter allows easier entry into the pulmonary artery, although it is more difficult to get across the tricuspid valve.

3. Using left femoral venous access, introduce a 5 Fr temporary pacing wire into the right ventricular apex (example settings: pacing 20 mA, sensing 3 mA). If cardiac output is poor, pacing may not be required to position the balloon at the valve. If venous access is difficult on the second attempt, an 0.014" wire with an IMA catheter can be placed as a landmark for the stick. Distal (the one that has writing on it) connects to the negative terminal on the pacemaker box.

4. Perform right heart catheterization. Important measurements include pulmonary artery pressure, pulmonary artery wedge pressure, and cardiac output/cardiac index by thermodilution

and Fick methods. Technically, you should wait until you have measured the aortic valve gradient.

5. At this point intravenous heparin should be administered to achieve a target ACT of >250 seconds.

6. Cross the aortic valve:

   i.   In the LAO straight projection, introduce an AL 1 catheter up to the aortic valve, in conjunction with a 0.035" straight wire (take this catheter over the arch with a regular J wire). Note the arterial waveform (for future assessment of possible aortic insufficiency), diastolic blood pressure, and gradient between the ascending aorta and femoral artery.

   ii. Cine the valve in LAO and RAO to observe the location of the orifice.

   iii. If crossing is difficult, alternate catheters (JR 4, MPA 2, AL 2, AR 1) or wires (Terumo straight glide wire, Wholey, or 0.014" Grand Slam) may help.

   iv. Often, an abnormal aorta or bicuspid valve creates an orientation for the valve that is very difficult to cross (if the valve cannot be crossed retrograde, consider antegrade approach).

   v.   Clock to move toward the left (analogous to clocking to get to the left circumflex artery from the left anterior descending artery). Keep the wire out of the catheter slightly to facilitate crossing of the valve. Cross the aortic valve in LAO (the danger in crossing in RAO is that the wire might injure the left main trunk without it being evident as you are viewing it in RAO).

   vi. If the AL 1 does not track all the way into the LV over the straight wire, but its tip is within the LV, remove the straight wire, place a long Wholey wire in the LV, and attempt to advance a 4 Fr glide catheter into the LV.

   vii. Place the Lunderquist 0.035" wire through the AL 1 in RAO and prepare to dilate the valve in RAO under cine. In patients with poor ventricular function, do not advance the wire all the way to the apex as these patients have a higher probability of having an apical thrombus.

    viii. If the Lunderquist wire does not sit well after it has been advanced to LV over the AL 1, reform it with the aid of a pigtail catheter (remove AL 1 first with the wire in the LV).

7. Assess aortic stenosis:

    i. Simultaneously record the LV and right femoral artery (RFA) pressures; first, record the ascending aorta and femoral artery simultaneously to see if there is any pressure difference. The pigtail catheter provides a more accurate measurement of LV pressure than the AL 1 catheter, which often dampens. Normally, the peak and mean gradients are equal up to 40 mmHg. This is because the "peak" is actually a peak-to-peak gradient (and not an instantaneous peak gradient from echo), which correlates fairly well with the mean gradient up to 40 mmHg; the invasive mean gradient correlates well with the mean echo gradient.

    ii. Another alternative for obtaining LV and Ao pressures is a dual lumen pigtail catheter, as discussed in the previous section. A dual lumen is often inaccurate as it can overestimate the gradient. Therefore, measure the gradient with the sheath pressure. The dual lumen pigtail catheter may occasionally come apart because of the two lumens disconnecting.

    iii. If there is concern about the severity in the setting of low ejection fraction, infuse dobutamine 10, 20, 40 µg/kg/min (or pace) to increase cardiac output (easier to measure by thermodilution method for these sets of measurements), and observe gradients and the aortic valve area.

    iv. To attempt to evaluate for dynamic outflow obstruction (hypertrophic cardiomyopathy physiology), one can manipulate the dual lumen pigtail catheter within the LV to induce PVCs and observe for Brockenbrough–Braunwald–Morrow sign (post-extra systolic potentiation of the gradient both by increasing LV systolic pressure and by lowering systemic systolic blood pressure). A midcavity gradient is not associated with a Brockenbrough–Braunwald–Morrow sign. To measure an intraventricular gradient, you can leave a 0.035" Lunderquist in, and place an 8

Fr MP guide over it, thereby sliding it in and out to estimate the location of the gradient (both intraventricular and valvular).

8. Prepare the balloon within the plastic casing. To do this, inflate the balloon using contrast/saline mix 30/70 cc, then suck back vigorously with a 20-cc syringe and stopcock; repeat five or six times. Flush the central lumen with saline. It is especially important to get all air out from larger balloons, as they are more likely to rupture. Laplace's law dictates that larger diameter generates more radial force and greater risk of balloon rupture.

9. Sizing of the balloon:
   i.   This is usually done based on measurements of the annulus obtained by echocardiography or CT scan. Most adults have an annulus size of 20–22 mm.
   ii. The most commonly used balloon is Z-MED II (range 4–30 mm; 6-cm length). A 12 Fr sheath accommodates up to 24 mm balloon withdrawal. For a 25-mm balloon a 14 Fr sheath is needed.
   iii. Usual practice is to start with a 20- or 22-mm balloon and change to a 23 mm balloon to halve the gradient or reach an aortic valve area of 0.9–1.0 cm^2. If just one balloon size is to be used, 23 mm is often a good selection. Occasionally a 24- or 25-mm balloon is also required (LV outflow tract diameter greater than 20 mm is reassuring in terms of going with 25-mm balloon; larger balloons such as 25-mm carry a greater risk of severe aortic insufficiency with bicuspid valves).
   iv. When the gradient is very high, it is prudent to make the initial inflation with a smaller balloon, such as 20 mm.
   v.   Anecdotally, chronic steroid use increases the risk of tearing the valve with balloon dilation.

10. Balloon dilation of the aortic valve (Figure 6.1):
   i.   Straddle valve with balloon, pace at 180–200 bpm (if arterioventricular conduction is not adequate, high pacing rate may create 2:1 block and half the rate; in that case lower the pacing rate to 180 or 160 bpm), inflate for 3–5 seconds (fast inflation is important to ensure that the balloon does not slip

Percutaneous balloon valvuloplasty

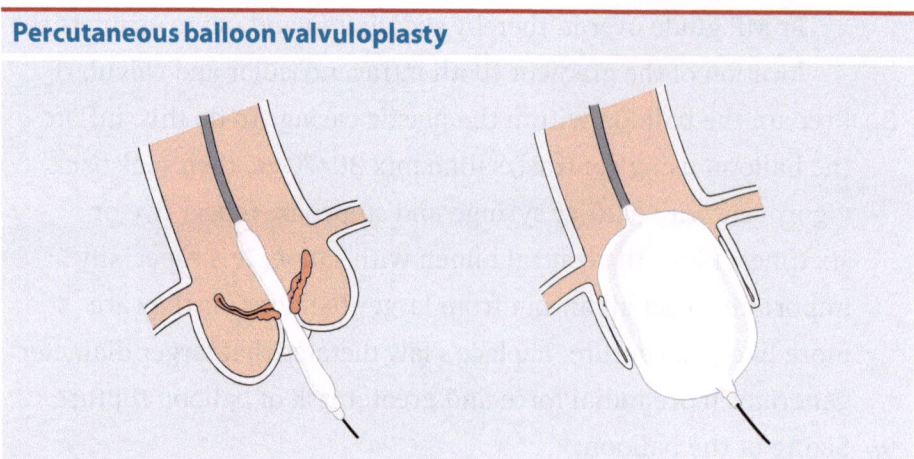

Figure 6.1 Percutaneous balloon valvuloplasty.

    out; if this happens the balloon must be immediately deflated), deflate and pull balloon back into ascending aorta (to reinsert, apply negative pressure as you advance it to the straddling position), and stop pacing.

  ii. When inflating, hook up a 60-cc syringe filled with contrast mix (30% contrast, 70% heparinized saline) to the stopcock-straight and endoflator (with heparinized saline) to the stopcock-side, and sequentially (syringe first, endoflator second to 6 atmospheres) inflate strongly.

  iii. If another inflation is planned, consider administering norepinephrine 8 µg as you are deflating the balloon (this allows for a quicker hemodynamic recovery).

  iv. If the balloon bursts, withdraw it to the edge of the sheath (do not pull it inside the sheath) while maintaining the Lunderquist wire position in the LV. Then remove the balloon/sheath combination while maintaining the Lunderquist wire position, cut the balloon with scissors, and reinsert the 12 Fr sheath with the dilator (there may be some bleeding around the sheath as the balloon may have stretched the arteriotomy).

11. Assess the result with the Fick cardiac output and cardiac index, LV-Ao/RFA gradient, and arterial waveform for aortic insufficiency (low diastolic blood pressure and sharp pressure

decline in early diastole; note that aortic insufficiency on the arterial pressure waveform may be difficult when the heart rate is fast and the diastole is short). A steeper upslope of the aortic waveform would indicate greater valve opening. To judge aortic regurgitation, diastolic pressure and hemodynamics (systolic blood pressure) are the best indicators (echo is not helpful; loss of dicrotic notch is not a reliable indicator of aortic insufficiency; and aortogram is complicated by the tenuous clinical status of these patients). In addition, make sure to auscultate the patients before (and after, if aortic insufficiency is suspected) balloon aortic valvuloplasty.

12. Hemostasis: tighten the previously deployed sutures from one of the Perclose devices as the sheath is removed and the wire is left in place. The second Perclose device is tightened after the wire is removed.

Hemodynamic distress

In cases of hemodynamic collapse (may be accompanied by seizures), support hemodynamics with norepinephrine 8 μg or 16 μg, or if needed, epinephrine, but 0.5 mg first (this may generate tachycardia and/or hypertension), especially if the patient is conscious. Be prepared to intubate, as severe pulmonary edema is common.

Bicuspid aortic valves may tear more easily creating severe aortic regurgitation. Severe aortic regurgitation is tolerated better, when pre-procedural left ventricular end diastolic pressure is low.

Considerations with a pre-existing intra-aortic balloon pump (IABP)

1. Removing an IABP cannot be done with closure devices.
2. Performing the valvuloplasty while on an IABP is not feasible given the inability to evaluate diastolic pressure.
3. The IABP cannot be removed while upsizing to a 12 Fr sheath because of the inability to place the Percloses (for the 12 Fr sheath) due to infection risk. The only way is to manually remove the IABP

prior to the valvuloplasty, or as you are performing it or right after balloon aortic valvuloplasty. Have the IABP on stand-by during the entire case, maintain a high ACT, be quick with the valvuloplasty, and remove the IABP immediately after the case.

Bicuspid valve

The strategy is the same for bicuspid aortic valve, but occasionally the nonfused leaflet, which is often less calcified than the fused leaflets, tears creating severe aortic regurgitation.

Prosthetic valves

Prosthetic valve valvuloplasty can cause catastrophic aortic insufficiency.

Pointers
- Beware of left ventricular endomyocardial perforation with end-hole catheter contrast injection (ST elevations, contrast hang-up, +/- perforation). If a tear occurs, anticoagulation to prevent a thrombus formation on the nidus of the tear may be indicated.
- When there is an ostial left main trunk stent, the Z-MED II balloon can make contact and injure the stent while being inflated.

Percutaneous aortic valve replacement

1. It is important to align the annulus perfectly as you deploy the valve. Perform ascending aortography to determine the angiographic angles, which visualize the true valve plane. That is, when the overlapped (in LAO, the right and non, and in RAO, the left and right) cusps have their inferior margins precisely superimposed, a line bisecting the inferior margins of the cusps is the valve plane. A good starting point for the views is LAO cranial 40/20 and RAO caudal 20/20.

2. In addition, use the biplane ascending aortogram to determine which leaflets and commissures are calcified and restricted, and whether the ascending aortic slope is vertical or horizontal. The

latter makes device delivery difficult. Also assess the optimal angles for device deployment: see the annulus as perpendicular as possible by lining up the inferior margins of the three cusps.

3. Use the ascending aortogram to measure the distance to the left main trunk and right coronary artery (you can also make an estimation of this with CT), especially if there is heavy calcification. Perform a pigtail ascending aortic injection during balloon predilation and inflation to judge the distance between the valve leaflet and left main trunk. CT has the additional advantage of allowing the assessment of aortic atherosclerosis and iliac caliber, tortuosity, or calcification. CT can also be used to determine the optimal angles for visualization of the true valve plane, although this is better archieved with angiography.

4. Transesophageal echocardiography is important for measuring annulus size and then for predeployment positioning of the device.

5. The skirt in the Edwards-Sapien percutaneous valve prevents paravalvular aortic regurgitation, as it is very difficult to be exactly at the annulus. The mechanism for aortic regurgitation is paravalvular if the valve is underdeployed and not fully opposing the annulus. The skirt below the valve plane is covered and does not allow for aortic regurgitation. Deploying so low that aortic regurgitation could occur above the valve is not possible, as it would cause the prosthesis to slip into the LV. If the valve stent is dilated too much with deployment, it can result in central aortic regurgitation (overstretched valve).

6. The Edwards-Sapien device for annulus >21 mm: 24 Fr (in reality = 26 Fr) access and 26-mm valve, and the iliac diameter required is 9 or 10 mm. For annulus <21 mm: 22 Fr (in reality = 24 Fr) and 23 mm, and the iliac diameter required is 7 or 8 mm. The Edwards-Sapien valve is 16 mm long and results in aortic valve area 1.6 cm^2 (the current CoreValve® area is 1.2 cm^2).

7. The smallest of the next generation of transcather aortic valve devices is 18 Fr.

8. In the PARTNER (Placement of AoRTic TraNscatherER valve) trial, IVUS of the iliac arteries (over a 0.014" wire from the short sheath) is important to allow for delivery of the prosthesis (10-mm common

iliac artery, 8-mm external iliac artery). If no guide is available, then bony landmarks should be used as dye cannot be injected. Even if caliber is adequate, calcification or tortuosity, which can be determined with fluoroscopy/descending aortography with peripheral runoffs, may be prohibitive. IVUS can also examine the vessel when it is still within the sheath (e.g., location with respect to a common femoral artery bifurcation). You can even see a just-deployed Perclose with the IVUS (disrupted intima with rolled-up area of the suture).

9. To angiographically visualize the aortic valve annulus level, use as a starting point RAO caudal 20/20, LAO cranial 40/30, and adjust as needed. Descending aortography allows evaluation of the descending aortic and iliofemoral tortuosity and sizing. In cases of tortuous iliacs, you can place a Lunderquist or stiff Amplatz wire to see if the artery straightens out adequately.

Procedure

1. The patient is intubated and general anesthetic is administered.
2. Transesophageal echocardiography is performed. Swan-Ganz catheterization, as described for aortic valvuloplasty, is performed through an 8 Fr venous sheath. A 5 Fr venous sheath is inserted for the temporary venous pacer. Arterial access, as described for valvuloplasty, is obtained, except that the sheath size is considerably larger (22 or 24 Fr), and therefore a surgical cutdown is often necessary. Two perpendicular 10 Fr Prostar devices are necessary for hemostasis.
3. Place a pigtail catheter in the aortic root for injections to evaluate the distance between leaflets and left main trunk during balloon valvuloplasty and to evaluate aortic regurgitation following device deployment (settings: 0/20/20/800; LAO cranial 40/30; RAO caudal 30/20). Observe the distance to the coronaries. Of note, a calcific nodule may be pushed to obstruct the left main trunk; observe leaflet nodules on echo during balloon valvuloplasty preceding the device deployment.

4. Place a super stiff Amplatz wire and perform balloon valvuloplasty with a 20 mm x 3 cm Z-Med-II balloon (3-cm balloons slip more easily than 5- or 6-cm balloons, but mimic the device deployment more closely as the device is on a 3-cm balloon). If needed, use a 22 mm x 5 cm Z-Med-II balloon. Hold ventilations during balloon valvuloplasty and during device deployment.

5. In addition to an arteriotomy, dilate the groin with a 24 Fr Cook dilator, followed by a 24 Fr Edwards dilator. Advance the prosthesis with the aid of the Flex sheath and a pusher. Mineral oil can be used to lubricate the device as a last resort. A cone-shaped tip balloon is currently under investigation to lower the crossing profile (this will not lower the crossing profile of the stent, which is on the balloon without any "encasing," but rather as a bare structure). Another feature impairing the ability to cross is wire bias to one of the commissures and away from the center of the valve orifice. Once the stent is out of the sheath, it cannot be withdrawn back inside, but has to be deployed. If it cannot be deployed at the aortic valve annulus, it has to be deployed elsewhere in the descending aorta.

6. Position and deploy 23- or 26-mm device while pacing and holding ventilations and injecting through the pigtail catheter in the ascending aorta (position both by angiography and transesophageal echocardiography). Deployment position: 80% of the length within LV and 20% of length within aorta. For an apical approach, it is important to generate a straight trajectory (50% of the length within LV and 50% within aorta); do not err on the LV side.

7. Deflate the balloon fully before you stop pacing, so that the ventricular contraction does not eject the stent while the balloon is still partially inflated. This results in a rather long period of pacing (considerably longer than with balloon valvuloplasty). The stent can be postdilated once (increasing central aortic regurgitation, but reducing periprosthetic aortic regurgitation). The area near the trigone, in cases of severe mitral annular calcification, is difficult to expand with the stent, and thus results

in aortic regurgitation (at the junction of the non and left cusps). For a larger annulus, use a 26-mm device.

8. After deployment, assess for aortic regurgitation with transesophageal echocardiography and with ascending aortogram (using a pigtail catheter). To prevent blood loss during Prostar deployment, advance a 9 x 20 mm OPTA balloon to ipsilateral common iliac artery and inflate at 4 atmospheres.

Pointers

- The antegrade approach is currently not approved by the FDA; however, it appears that this approach has no iliofemoral arterial issues, so it may be easier to cross. The resultant iatrogenic atrial septal defect (ASD) is not a problem. Maintaining a wire loop within the LV is important to prevent injury to the anterior mitral leaflet.

- For the apical approach, it is very important to generate a straight trajectory and the best starting point for this is intercostal spaces 5 and 6. The device must be inserted in the correct orientation (opposite of the femoral approach).

- If one of the prosthesis cusps is locked open, a second prosthesis should be placed within it.

See Figure 6.2 for a transcather aortic valve implantation (TAVI).

Patent foramen ovale/atrial septal defect

Indication

Closure of patent foramen ovale (PFO) is controversial, as enrollment into randomized clinical trials has been slow. There is no on-label indication in the United States for PFO closure. The CLOSURE I trial also failed to show a reduction in recurrent stroke or transient ischemic attack with PFO closure in patients with cryptogenic stroke (see Recommended Reading). The Migraine Intervention with STARFlex Technology (MIST) trial failed to show a reduction in the incidence of migraine with PFO closure in patients with cryptogenic stroke (see Recommended Reading). The most stringent indication (still off-label) for PFO closure falls under the FDA's Humanitarian Device Exemption (HDE). The HDE was created

Transcatheter aortic valve implantation (TAVI)

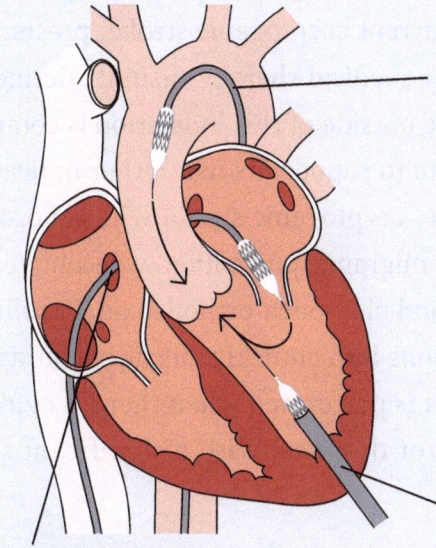

Retrograde or transfemoral technique
The catheter is advanced to the stenotic aortic valve via the femoral artery

Advantages
- Faster, technically easier than antegrade approach

Disadvantages
- Potential for injury to the aortofemoral vessels
- Crossing the stenotic aortic valve can be challenging

Antegrade technique
The catheter is advanced via the femoral vein, traversing the interabial septum and the mitral valve, and is positioned within the diseased aortic valve

Advantages
- Femoral vein accommodates the large catheter sheath
- Easy management of peripheral access site

Disadvantages
- Risk of mitral valve injury and severe mitral valve regungitation
- Correctly positioning the prosthetic valve can be challenging

This technique is no longer in use

Transapical technique
A valve delivery system is inserted via a small intercostal incision. The apex of the LV is punctured, and the prosthetic valve is positioned within the stenotic aortic valve

Advantages
- Access to the stenotic valve is more direct
- Avoids potential complications of a large peripheral access site

Disadvantages
- Potential for complications related to puncture of the LV
- Requires general anesthesia and chest tubes

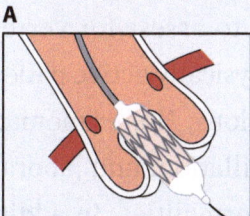 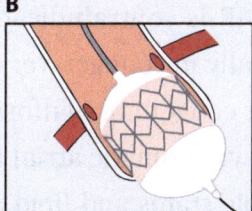 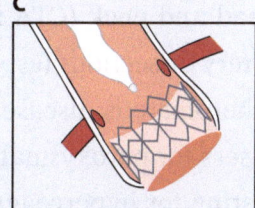

The aortic valve prosthesis is placed at mid-position in the patient's aortic valve so as not to impinge on the coronary ostia or to impede the motion of the anterior mitral leaflet (A). The prosthesis is deployed by inflating (B), rapidly deflating, and quickly withdrawing the delivery ballon (C)

Figure 6.2 Transcatheter aortic valve implantation (TAVI). LV, left ventricle. Reprinted with permission from Singh IM, et al. *Cleve Clin J Med.* 2008;75:805-12.

by Congress to streamline the approval process for devices that would serve fewer than 4,000 patients in the United States annually. The HDE indication for PFO closure is recurrent cryptogenic stroke, presumably due to a PFO, despite appropriate medical therapy (usually defined as anticoagulation therapy). Closure outside of this indication is common, although there is a paucity of data to support its use. Other indications include a first cryptogenic stroke, cryptogenic stroke with no or inadequate medical therapy, recurrent migraine, prevention of decompression sickness among deep sea divers, and platypnea-orthodeoxia. Enrollment of patients with off-label indications into clinical trials is encouraged.

Closure of atrial septal defect is performed when there is evidence of right-sided heart failure and/or dyspnea, with a significant shunt fraction.

Imaging

Transesophageal echocardiography is useful in outlining the anatomy of most PFOs.

Testing

Since PFO prevalence is estimated to be 25–30% of the global population, merely closing a PFO in a patient with cryptogenic stroke may not prevent recurrent strokes if other factors are responsible. Hence, in all patients with cryptogenic stroke where PFO closure is being contemplated, it is important that a detailed investigation for other causes is conducted. These include, but are not restricted to: MRI/MRA of the head and neck (CTA if MR is contraindicated) to assess for vertebral artery dissections (especially in younger, very physically active patients), atheromatous disease, or congenital malformations; Holter monitor to assess for paroxysmal, asymptomatic atrial fibrillation; and laboratory testing for hypercoaguable states and lipid abnormalities. In addition, age-appropriate cancer screening should be considered.

Anticoagulation

Unfractionated heparin to achieve an ACT of >250 seconds.

Access

Basic setup consists of a short 8 Fr sheath in the right femoral vein, which will be upsized later with the delivery sheath depending on what device is used. A 35-cm 8 Fr sheath is inserted in the left femoral vein, which will be used for intracardiac echocardiography (ICE). In the straight projection, the echo (ICE) should point to the patient's left, while in the lateral projection it should point anterior.

Procedure

There are six basic steps for closure of an ASD:

1. Pulmonary angiogram to evaluate for anomalous pulmonary venous return.
2. Obtain right-sided pressures and saturations.
3. Cross the atrial defect with a soft 0.035" wire.
4. Exchange the soft wire for a 1 cm Amplatz super stiff wire through a Goodale-Lubin (GL) catheter.
5. Size the defect with a sizing balloon.
6. Deploy the closure device. The first two steps are not necessary for closure of a PFO.

These are described in more detail below.

1. Pulmonary angiogram is performed with a Berman angiocatheter, NIH catheter, or pigtail catheter using a rate of 20–40 cc/sec for a total volume of 40 cc. The NIH catheter may be difficult to position. This can be facilitated by confirming anterior direction in the lateral projection when the catheter is in the right ventricle. The Berman catheter is the easiest to manipulate and also likely to be the safest, since it is balloon tipped. The Berman wedge is railed over a 0.035" wire, is balloon tipped, and has an end hole, but cannot be used to perform power injection. The Berman angiocatheter tracks are balloon tipped; however, there are just side holes for power injection. Since there is no end hole, this catheter does not track over a wire. Once the catheter is in place, power injection is performed. Cine is continued until the *levo* phase to confirm that four pulmonary veins drain into the left atrium and that the left atrium fills prior to right atrial filling to exclude anomalous pulmonary venous return.

Determination of shunt fraction

$$Qp/Qs = \frac{\text{Arterial saturation} - \text{Mixed venous saturation}}{\text{Pulmonary venous saturation} - \text{Pulmonary artery saturation}}$$

In the absence of pulmonary venous saturation, arterial saturation can be used instead.
For example:

$$Qp/Qs = \frac{97-76}{97-85} \qquad\qquad Qp/Qs = 1.8$$

Figure 6.3 Determination of shunt fraction. <1.5 = small; 1.5 to 2.0 = medium; >2.0 = large.

Suspected anomalous pulmonary venous return can be confirmed by performing a selective angiogram of the anomalous vein into the superior vena cava (SVC).

2. Pulmonary artery saturations and pressures are obtained with one of the above listed catheters. The GL catheter is used to sample saturations and pressures in the high SVC, low SVC, right atrium, and IVC. Saturations are used to determine the Qp/Qs ratio (Figure 6.3).

3. A J-tip 0.035" wire and a GL/MP catheter is used to cross the PFO or ASD into the left superior or inferior pulmonary vein. Placement of the wire beyond the cardiac silhouette helps confirm that the wire is in a pulmonary vein and not in the left atrial appendage. Wire placement is further confirmed in the lateral projection; anterior direction of the wire suggests placement in the left atrial appendage. If there is difficulty engaging a pulmonary vein, a soft-tipped 0.035" wire can be used to provide more direction. Once the wire is in a pulmonary vein, the GL catheter is advanced to the ostium of a pulmonary vein.

4. The J-tip wire is now removed and a 1-cm super-stiff Amplatz wire is crossed into the pulmonary vein. The Amplatz wire provides the rail to deliver the sizing balloon and device.

5. A sizing balloon is used to cross the PFO or ASD (sizes are 25, 30, and 35 mm NMT or 24, 28, 32, and 40 mm Amplatzer). The balloon is calibrated (10-cm markings on the NMT and 15-cm

markings on the Amplatzer), inflated across the defect, the size of the defect measured, and then the balloon is removed.

6. The final step requires preparing the delivery sheath on the back table. The basic setup consists of a long delivery sheath with dilator, the cable that attaches to the closure device, and the device introducer sheath. First, insert the flushed dilator into the long flushed delivery sheath. Then attach the device to the delivery cable and advance the delivery sheath on the cable. The device is screwed on the cable (clockwise 5 times until it clicks), submerged under sterile saline (1 g vancomycin can be added), and withdrawn into the introducer sheath as it is flushed. Next, the right femoral vein short sheath is removed (keeping the super-stiff Amplatz wire in place) and the long delivery sheath/dilator is advanced into the left atrium where the dilator is broken. Stopping the dilator at this point prevents the stiff tip from damaging the roof of the left atrium. A gentle injection through the sheath is performed under fluoroscopy (from the groin to the left atrium) to make sure there is no entrained air in the sheath. At this point the Amplatz wire is removed and the cable with device introducer sheath is inserted into the delivery sheath. The cable is advanced approximately 20–30 cm, at which point the device introducer sheath is removed. The device is delivered to the distal portion of the delivery sheath. When ready to deploy, the device is advanced out of the delivery sheath until the left atrial disk is fully expanded in the left atrium (sizes of Amplatzer occluders: 4–20 mm [in 1-mm increments] and 22–38 mm [in 2-mm increments]) The left atrial disk is pulled back firmly against the interatrial septum using fluoroscopy and ICE. With the cable anchored in place, the delivery sheath is pulled back to the right atrium to unsheath the right atrial disk. The Minnesota wiggle gently pushes and pulls the device under fluoroscopy to make sure it does not pull though the defect. For deployment, the cable is rotated counter clockwise four times, then under fluoroscopy (or cine) it is rotated a final time. The cable has a tendency to stab forward when it is disconnected and should, therefore, be promptly pulled back upon disconnection.

Caveats

A long PFO tunnel may necessitate placing a device across a trans-septal puncture. See the section on trans-septal punctures (page 83). Gentle pushing and pulling of this device after the left and right atrial disks have been deployed is necessary to make sure the right atrial disk is flattened against the septum.

Alcohol septal ablation

Indication

Refractory symptomatic hypertrophic obstructive cardiomyopathy (HOCM) in a patient who is not a surgical candidate.

Access

A 6 Fr sheath in the right internal jugular for screw-in pacer, a 4 Fr sheath in femoral artery for pigtail, and at least a 7 Fr sheath in the femoral artery for guide catheter.

Equipment

Short over-the-wire balloon: 1.5/9, 2.0/9, or 2.5/9 mm, along with a 300-cm 0.014" wire (Prowater or Balance Middleweight).

Anticoagulation

Heparin.

Procedure

1. Administer heparin to achieve a high ACT (i.e., >300 seconds). A larger guide size is used to prevent coronary thrombosis occurring due to a long procedural time.
2. Advance the pigtail catheter in the LV and measure simultaneous LV and Ao gradients. PVC can be induced with pigtail catheter to demonstrate the Brockenbrough-Braunwald-Morrow phenomenon. Obtain baseline transthoracic echocardiographic images (apical 4 or 3 chamber).
3. In a posterior–anterior cranial projection, wire the target septal perforator. Advance an over-the-wire balloon to the proximal septal perforator. The proximal tip of the balloon may be in the left

anterior descending artery to ensure that the maximal length of septal perforator is sclerosed.

4. Inflate the balloon at approximately 6–12 atmospheres depending on the balloon size. Observe pressures since this should reduce the gradient. Inject through the guide to confirm that there is no antegrade flow in the septal branch, and that there is still good flow in the LAD.

5. Remove the wire. Inject contrast through the over-the-wire balloon to confirm echocardiographic location of where the infarct will be.

6. Inject 0.2 cc of 96% anhydrous alcohol into the over-the-wire balloon. This much volume is needed to fill the dead space of the balloon.

Mitral balloon valvuloplasty

Indication

Mitral balloon valvuloplasty produces comparable and durable results to surgical commissurotomy and is considered the procedure of choice for rheumatic mitral stenosis. The mechanism of benefit is to open fused mitral commissures.

Work-up

Transthoracic and transesophageal echocardiography (TEE) is performed to define the mechanism and severity of mitral stenosis. Echo also assesses the degree of mitral regurgitation, valvular calcification, mobility, and thickening and subvalvular thickening or calcification. Lastly, the left atrial appendage must be evaluated for the presence of thrombus prior to valvuloplasty.

A split score can be performed where each variable receives a score of 0 to 4 (see Table 6.1). Valves with scores below 8 to 10 are usually considered suitable. Higher scores usually indicate the need for mitral valve surgery.

The gradient across the mitral valve is dependent on heart rate and filling pressures; therefore, provocative maneuvers (i.e., fluid, nitroglycerine) to increase the heart rate can be performed if there are equivocal hemodynamics.

Splitability score

1. Valvular thickening

- Near-normal (4–5 mm)
- Thickened tips
- Entire leaflet thickened (5–8 mm)
- Marked leaflet thickening (>8–10 mm)

2. Valvular calcification

- Single area of brightness
- Scattered areas at leaflet margins
- Extends into mid leaflets
- Extensive leaflet brightness

3. Leaflet mobility

- Highly mobile
- Decreased mobility at the tips
- Basal leaflet motion only
- Minimal to no motion

4. Subvalvular thickening/calcification

- Minimal chordal thickening
- Thickening of <1/3 of the chordal apparatus
- Distal 1/3 of the chordae thickened
- Extensive thickening of the papillary muscles

Table 6.1 Splitability score.

Anticoagulation

Heparin.

Access

1. Access begins with a short 8 Fr sheath in the right femoral vein, a short 8 Fr sheath in the left femoral vein (for Swan-Ganz catheter), and a 4 Fr sheath in the left femoral artery. An incision is made in the right groin (approximately 1.5 cm) and spread with a Kelly clamp to facilitate movement of the Inoue balloon.
2. Arterial access is needed for the following reasons:
    i. Placement of a pigtail catheter in the LV to measure simultaneous left arterial (LA) and LV pressures.

ii. Placement of a pigtail catheter in the ascending aorta to facilitate posterior orientation of the Brockenbrough needle during trans-septal puncture.

iii. To perform coronary angiogram if indicated. The left groin is preferred for arterial access to avoid close proximity to the frequent exchanges of the valvuloplasty equipment in the right groin.

3. At this point, the transesophageal echo is placed and the left atrial appendage evaluated if this was not previously done.

Intervention

1. Trans-septal puncture:

   i. Change the 8 Fr sheath in the right femoral vein to an 11 Fr Mullins sheath/dilator and advance to the SVC over a 0.032" wire.

   ii. Remove the wire and insert the Brockenbrough needle just proximal to the tip of the Mullins sheath/dilator. Attach the proximal end of the Brockenbrough needle to a pressure tracing. Hold the Mullins/Brockenbrough as a unit with the hub at "4 o'clock." Pull the Mullins sheath just below the fossa ovalis, but not to the coronary sinus. Use biplane to confirm correct alignment of the Mullins sheath.

   iii. In the lateral view, point the Mullins sheath posterior to the pigtail, which is in the noncoronary sinus. Transesophageal echocardiography is also used to facilitate puncture. A puncture that is too high in the septum will result in difficulty crossing the valve. Similarly, if a patent foramen ovale is crossed instead of puncturing the atrial septum, there may be difficulty in crossing the valve. When correct position of the Mullins sheath is confirmed, the septum is punctured with the Brockenbrough needle and the Mullins sheath advanced over the needle into the left atrium, at which point the needle is removed.

   iv. Confirm that the right atrial pressure has changed to left atrial pressure. A drop or no pressure could indicate puncture into the pericardial space, while an increase in pressure could indicate puncture in the aorta.

    v. Administer heparin to achieve a target ACT of >250 seconds.

    vi. At this point document LA and LV pressures, cardiac output and index, and mitral valve area before proceeding with mitral valvuloplasty.

2. The Inoue balloon has three ports: the vent port; which connects to the outer lumen (the longer tube); the balloon port, which connects to the inner lumen (the short tube); and the shaft (metal end).

    i. Flush the shaft with saline and the vent port with a mixture of 70/30 saline/flush. Flushing through the vent port fills the outer lumen, then the inner lumen, and out through the balloon port.

    ii. At this point the vent port is closed and capped since it will not be used again. Capping prevents inadvertent attachment to the vent port during intended balloon inflation.

    iii. The next step is to test the balloon for sizing. The Inoue balloon has three sizes based on the inflated volume of the 70/30 mix. To determine the maximum balloon size, take the patient's height in centimeters, divide by 10, then add 10. Most sizes are between 20 and 28 mm. Start 2–4 mm below maximum and increase at 2-mm increments to the maximum.

    iv. After determining the maximum size, insert the metal stylet into the metal hub of the Inoue balloon to stiffen the body of the balloon (metal to metal). Then attach the metal portion of the stylet to the plastic portion of shaft (metal to plastic via an L-type connection). This elongates the balloon, meaning it has a lower profile and can cross the septum.

3. Perform mitral valvuloplasty:

    i. Advance the Toray wire through the Mullins sheath and exchange the Mullins sheath for the 14 Fr black dilator. This dilates the venotomy and interatrial septum.

    ii. Advance the prepared Inoue balloon to the roof of the left atrium. Disconnect "metal to plastic" while advancing the balloon forward to approximately 6 or 7 o'clock. Disconnect "metal to metal" then remove the Toray wire/stylet.

iii. Insert the torqueable stylet and with a counterclockwise rotation, point the nose of the balloon into the LV.

iv. Advance the Inoue balloon over the stylet into the LV. Inflate the distal end of balloon and pull it back until it grabs/straddles the valve. Confirm the position is correct with transesophageal echocardiography and fluoroscopy.

v. Inflate the balloon fully for 3 seconds and deflate rapidly.

vi. Assess the degree of mitral regurgitation and LA–LV gradients. Success is defined as a decrease in gradient of more than 50% or mitral valve area of >1.5 cm². If success has not been achieved and there is no moderate or worse mitral regurgitation, repeat valvuloplasty with a larger balloon.

4. To remove the Inoue balloon, insert the Toray wire, then the metal stylet ("metal to metal" then "metal to plastic"), and withdraw to the right side of the body.

Miscellaneous

Pulmonary angiogram

Power injection can be performed through the NIH or Berman angio-catheter at 15–20 cc for 30–40 cc total contrast (see Table 2.5). The NIH catheter can be utilized to obtain adequate images of the main pulmonary artery and its proximal branches. However, if pulmonary angiography is performed for other reasons, such as for detection of pulmonary arteriovenous malformations, the stiffer Berman angiocatheter is usually necessary.

Recommended reading

Articles

Bates ER, Babb JD, Casey DE Jr, et al. ACCF/SCAI/SVMB/SIR/ASITN 2007 clinical expert consensus document on carotid stenting: a report of the American College of Cardiology Foundation Task Force on Clinical Expert Consensus Documents (ACCF/SCAI/SVMB/SIR/ASITN Clinical Expert Consensus Document Committee on Carotid Stenting). *J Am Coll Cardiol* 2007;49:126–70.

Coeytaux RR, Williams JW Jr, Gray RN, Wang A. Percutaneous heart valve replacement for aortic stenosis: state of the evidence. *Ann Intern Med* 2010;153:314–24.

Dowson A, Mullen MJ, Peatfield R, et al. Migraine Intervention With STARFlex Technology (MIST) trial: A prospective, multicenter, double-blind, sham-controlled trial to evaluate the effectiveness of patent foramen ovale closure with STARFlex septal repair implant to resolve refractory migraine headache. *Circulation* 2008;117:1397–404.

Furlan A, Humphrey A, CLOSURE I Trial Investigators. A prospective, multicenter, randomized controlled trial to evaluate the safety and efficacy of the STARFlex septal closure system versus best medical therapy in patients with a stroke or transient ischemic attack due to presumed paradoxical embolism through a patent foramen ovale. Presented at the American Heart Association Scientific Sessions, Chicago, IL, November 15, 2010.

Hirsch AT, Haskal ZJ, Hertzer NR, et al. ACC/AHA 2005 guidelines for the management of patients with peripheral arterial disease (lower extremity, renal, mesenteric, and abdominal aortic): executive summary a collaborative report from the American Association for Vascular Surgery/ Society for Vascular Surgery, Society for Cardiovascular Angiography and Interventions, Society for Vascular Medicine and Biology, Society of Interventional Radiology, and the ACC/AHA Task Force on Practice Guidelines (Writing Committee to Develop Guidelines for the Management of Patients With Peripheral Arterial Disease) endorsed by the American Association of Cardiovascular and Pulmonary Rehabilitation; National Heart, Lung, and Blood Institute; Society for Vascular Nursing; TransAtlantic Inter-Society Consensus; and Vascular Disease Foundation. *J Am Coll Cardiol* 2006;47:1239–312.

Kern MJ, Samady H. Current concepts of integrated coronary physiology in the catheterization laboratory. *J Am Coll Cardiol* 2010;55:173–85.

King SB 3rd, Smith SC Jr, Hirshfeld JW Jr, et al. 2007 focused update of the ACC/AHA/SCAI 2005 guideline update for percutaneous coronary intervention: a report of the American College of Cardiology/American Heart Association Task Force on Practice guidelines. *J Am Coll Cardiol* 2008;51:172–209.

Krishnaswamy A, Klein JP, Kapadia SR. Clinical cerebrovascular anatomy. *Catheter Cardiovasc Interv* 2010;75:530–9.

Kushner FG, Hand M, Smith SC Jr, et al. American College of Cardiology Foundation; American Heart Association Task Force on Practice Guidelines. 2009 focused updates: ACC/AHA guidelines for the management of patients with ST-elevation myocardial infarction (updating the 2004 guideline and 2007 focused update) and ACC/AHA/SCAI guidelines on percutaneous coronary intervention (updating the 2005 guideline and 2007 focused update): a report of the American College of Cardiology Foundation/American Heart Association Task Force on Practice Guidelines. *J Am Coll Cardiol* 2009;54:2205–41.

Layton KF, Kallmes DF, Lindell EP, Cox VS. Bovine aortic arch variant in humans: clarification of a common misnomer. *Am J Neuroradiol* 2006;27:1541–42.

Nicholls SJ, Sipahi I, Schoenhagen P, et al. Application of intravascular ultrasound in anti-atherosclerotic drug development. Nat Rev Drug Discov 2006;5(6):485–92.

Nobuyoshi M, Arita T, Shirai S, et al. Percutaneous balloon mitral valvuloplasty: a review. *Circulation* 2009;119:e211–9.

Salem DN, O'Gara PT, Madias C, Pauker SG; American College of Chest Physicians. Valvular and structural heart disease: American College of Chest Physicians Evidence-Based Clinical Practice Guidelines (8th Edition). *Chest* 2008;133:593S–629S.

Singh IM, Shishehbor MH, Christofferson RD, et al. Percutaneous treatment of aortic valve stenosis. *Cleve Clin J Med* 2008;75:805–12.

Vahanian A, Alfieri O, Al-Attar N, et al.; European Association of Cardio-Thoracic Surgery; European Society of Cardiology; European Association of Percutaneous Cardiovascular Interventions. Transcatheter valve implantation for patients with aortic stenosis: a position statement from the European Association of Cardio-Thoracic Surgery (EACTS) and the European Society of Cardiology (ESC), in collaboration with the European Association of Percutaneous Cardiovascular Interventions (EAPCI). *Eur Heart J* 2008;29:1463–70.

Books

Aviles RJ, Messerli AW, Askari AT, Penn MS, Topol EJ (eds.). *Introductory Guide to Cardiac Catheterization.* Philadelphia, PA: Lippincott Williams & Wilkins, 2004.

Bhatt DL (ed.). *Guide to Peripheral and Cerebrovascular Intervention.* London, UK: Remedica Publishing, 2004.

Casserly IP, Sachar R, Yadav JS (eds.). *Manual of Peripheral Intervention.* Philadelphia, PA: Lippincott Williams & Wilkins, 2005.

Ellis SG, Holmes DR (eds.). *Strategic Approaches in Coronary Intervention.* Philadelphia, PA: Lippincott Williams & Wilkins, 2006.

Topol EJ (ed.). *Textbook of Interventional Cardiology.* Philadelphia, PA: Elsevier, 2003.